Hydropower: Renewable Energy Essentials

Hydropower: Renewable Energy Essentials

Jeff Caldwell

Larsen & Keller
www.larsen-keller.com

Hydropower: Renewable Energy Essentials
Jeff Caldwell
ISBN: 978-1-64172-077-9 (Hardback)

⊟ Larsen & Keller

Published by Larsen and Keller Education,
5 Penn Plaza,
19th Floor,
New York, NY 10001, USA

Cataloging-in-Publication Data

Hydropower : renewable energy essentials / Jeff Caldwell.
    p. cm.
Includes bibliographical references and index.
ISBN 978-1-64172-077-9
1. Water-power. 2. Renewable energy sources. 3. Water resources development.
I. Caldwell, Jeff.
TC147 .H93 2019
627--dc23

For more information regarding Larsen and Keller Education and its products, please visit the publisher's website www.larsen-keller.com

Table of Contents

Preface

Hydropower is the power that is harnessed from the energy of water that is falling or fast running. It is a form of renewable energy source that is used for irrigation, for operating mechanical devices like textile mills, sawmills, domestic lifts and ore mills. It is also used for generating electricity. Hydropower projects can be of various types, such as small hydro, micro hydro, conduit hydroelectricity projects, conventional hydroelectric, pumped-storage hydroelectricity, etc. This book unfolds the innovative aspects of hydropower and hydropower technologies, which will be crucial for the holistic understanding of the subject matter. It studies, analyzes and upholds the pillars of hydropower and its utmost significance in modern times. The book is appropriate for those seeking detailed information in this area.

A foreword of all Chapters of the book is provided below:

Chapter 1, Hydropower is the power that is harvested from the energy of fast running or falling water. This is an introductory chapter which will introduce briefly all the significant aspects of hydropower; **Chapter 2**, Hydroelectricity is the form of electricity produced using hydropower. The diverse aspects of hydroelectricity such as head and flow, hydropower potential, tidal energy, wave power, etc. have been thoroughly discussed in this chapter; **Chapter 3**, A rotary machine through which kinetic and potential energy of water is converted into mechanical work is called a water turbine. These can be categorized into two groups, namely, impulse turbine and reaction turbine. This chapter closely examines the diverse aspects of wind turbines and their role in hydropower generation; **Chapter 4**, A watermill is a structure which employs a water wheel or turbine to drive mechanical processes. This chapter has been carefully written to provide an easy understating of the different kinds of watermills such as tide mills, sawmills, gristmills, etc. and their role in hydropower generation; **Chapter 5**, A number of technologies and techniques are used in generating hydropower. This chapter discusses and analyzes energy generation through dams and the concepts of ocean thermal energy conversion, tidal barrage, tidal current, etc. in detail; **Chapter 6**, Hydropower is considered a means for economic development. Power generated using hydropower has a number of different applications. This chapter explores the use of hydroelectricity for flood risk management, in irrigation, business, etc.

I would like to thank the entire editorial team who made sincere efforts for this book and my family who supported me in my efforts of working on this book. I take this opportunity to thank all those who have been a guiding force throughout my life.

Jeff Caldwell

Introduction to Hydropower

Hydropower is the power that is harvested from the energy of fast running or falling water. This is an introductory chapter which will introduce briefly all the significant aspects of hydropower.

Water Resource

Water resources typically are supplies of water that could be beneficial to people. Making the distinction between water and water resources is important because not all of the water to which humans have access is suitable for their needs. Humans generally need fresh water, and most of the water on Earth does not fall into that category. Sources of fresh water include lakes, rivers, and ice caps.

Potential water resources for humans are limited when compared with the total amount of water on Earth. Only three percent of water on the Earth is fresh water. Although fresh water is distributed across the globe, the majority of it is frozen in the form of glaciers and the polar ice caps. This leaves most people on Earth to rely on rivers, lakes, and groundwater. These water resources are becoming subjects of increasing problems.

At one time, there was not a great deal of attention paid to water resources for humans. In modern times, however, people are beginning to realize that the subject cannot be ignored. Emphasis on conserving and protecting those resources has grown into international efforts. This has happened because some places already are experiencing water shortages. Experts foresee problems for much larger portions of the population in the future if prevention measures are not taken.

There are a number of things that are responsible for the present and impending water problems. Pollution, climate change, and urbanization are a few of the common examples. Climate change is affecting resources by causing rivers and lakes to dry up. Many of these water resources have been used for centuries to support household and agricultural uses. There often are not any sustainable alternatives readily available.

Urbanization affects water resources because many cities are growing but their water supplies are not. More people living in one place means that the water requirements in that place are greater. Unfortunately, it is not an option to make limitless supplies of water available to them.

Pollution involves people dirtying their water resources. Sometimes this is done by individuals and sometimes it is done by industrial entities. Dealing with pollution often is not simple. In some cases, even when it is technically possible, there is a lack of resources to do so.

Saltwater sources can in some cases be used as water resources. This is done by way of a process known as desalination, which involves removing the salt from the water. This is an expensive process and it is not widely used. It is not generally promoted as a solution for water problems.

Much effort is being invested by governments and civil society organizations to make people aware of the issues surrounding water resources for humans. Tips and solutions often are offered to educate people on how to change their habits. It is generally believed that sustainable solutions must include the efforts of all levels of society.

Types of Water Resources

Atmospheric Water

Atmospheric water, transforming from one state to another, participates in water circulation in nature. Moisture is mainly in a gaseous state in the atmosphere. The average content of water vapor decreases with height and latitude and depends on the season and type of underlying surface. Although it has only a relatively small moisture content, the atmosphere is the only source of fresh water regeneration in nature (through evaporation) and the main means of replenishment of water reserves (through precipitation). The total evaporation from the ocean surface and continents amounts to 577,000 km3 per year (a mean water layer of 1.13 m over the Earth); it consumes on average 88 W/m2 of heat, which amounts to more than a third of the solar energy supply of the Earth. Through the operation of the water circulation on the Earth, the entire 577,000 km3 falls to the Earth each year. Meridional water vapor transfer is a significant peculiarity of water circulation.

Water vapor (H_2O) is the most variable of the atmosphere components. Its volumetric content can change by a factor of 100,000, depending on the season and place. The lowest water values are observed either at a considerable height in the atmosphere or above the Antarctic plateau, and the maximum observed in subtropical and equatorial areas, where it often amounts to more than 3–4 percent. To determine water vapor content in a given air volume (the part of total atmospheric pressure caused by water vapor), a notional water vapor pressure (partial pressure) is used.

Alternatively, specific humidity can be used. This is the ratio of water vapor mass to the humid air mass. Absolute humidity is the ratio of water vapor mass to the humid air volume. The percentage ratio of water vapor partial pressure to the pressure of the water vapor saturated under the given temperature is called relative humidity.

Water in the atmosphere mainly comes from evaporation, from the surface of both the ocean and the land. Transpiration and direct evaporation from the surfaces of green plant (evapotranspiration), and evaporation from ice (sublimation) also increase the water content of the atmosphere. Condensation occurs when air, being saturated with water vapor, becomes cooler, and relative humidity reaches 100 percent. Condensation can be in the form of dew or rime. Dew forms when the air temperature is equal to the temperature of condensation and is higher than the temperature of any surface. Rime occurs if the temperature of the air and the surface is less than zero.

Condensation in the atmosphere occurs on condensation nuclei: small particles of dust, salt crystals, smoke, and so on, and results in cloud formation. Clouds are classified into types, kinds, and varieties according to their form and state (water drops or ice crystals). Depending on the synoptical and thermodynamical conditions of precipitation, clouds are divided into the following main types: total, shower, and mixed. The occurence of total precipitation is connected to a large-scale vertical air motion caused by a frontal or orographic rise, or large-scale horizontal convergence. Shower precipitations are formed in cumulus-rainy clouds under conditions of mesoscale convection in the unstable air of the layer within 5–10 km. Clouds mixed as to their genesis are caused by the simultaneous effect of regulated and convection motion, whose input differs depending on the precipitation type.

Fog is an accumulation of the smallest water drops or ice crystals decreasing visibility in and near the surface air layer. There are fogs of cooling and evaporation. Radiation fog can be formed when there are no clouds and the land surface is cooled by long-wave heat radiation at night. Advective fog can be formed by the horizontal movement of a relatively warm air mass over a colder air mass lying over the land surface.

Oceans, Inland Seas, Costal Zones and Estuaries

The seas are classified as interior (inland), marginal, or inter-insular according to their geographic position and degree of isolation. An inland sea is one almost completely surrounded by land and joining the ocean or adjacent seas only through relatively narrow channels. Interior seas are assumed to be subdivided into continental and intercontinental types. A continental sea is usually shallow, deeply intruding into the land within a continent (e.g. the Azov, the White and Black Seas, and Hudson Bay). An intercontinental sea is a part of the World Ocean located between continents and connected with the ocean or other seas by channels (e.g. the Mediterranean and the Red Sea). A marginal or adjacent sea is a part of the World Ocean adjoining the continent and partially separated from it by peninsulas or a group of islands, or simply by the ocean bottom uplifting. Marginal seas can be on the continental shelf (a shelf sea) or on the continental slope (e.g. the Barents, the Laptev, and the Norwegian Seas). An interinsular sea (encircled by islands) is a part of the World Ocean surrounded by a more or less dense circle of islands, the straits between which prevent a free water exchange with an open ocean (e.g. the Japan, and Sulu Seas).

The chemical composition of seawater, particularly in marginal and continental seas, can differ considerably from that of ocean water. This is because it is affected by the run-off from the land. An abundant runoff of rivers and springs (surface and ground-water discharges) changes marine water content considerably. The salinity of water in oceans and marginal seas is also very much affected by evaporation. However the chemical composition of seawater is practically constant (the exception being in zones of river water entry).

Water in the coastal zone is very mixed, being affected by waves and accompanying currents. The marine boundary of this zone is determined by near bottom wave veloc-ities and is limited by depths equal to ten times the wave height. An estuary is a par-tially enclosed coastal water area freely connected with the ocean, within which the seawater is considerably diluted by freshwater flowing from a river catchment area. These are transient zones between the land and sea. They can be subdivided into coastal plain estuaries, fiords, flooded river valleys (Chesapeake Bay), estuaries with coastal barriers (Pamlico Sound), estuaries of tectonic origin (San Francisco Gulf), and deltas. All estuaries are subdivided into three classes: "positive" where seawater is diluted by freshwater (river runoff with precipitation exceeding water losses for evaporation); "reverse" estuaries with increased salinity, where losses from evapo-ration exceed discharge and precipitation; and "neutral" where evaporation is fully compensated by discharge and precipitation. Estuaries can also be subdivided in re-lation to the way in which mixing takes place. They are usually subdivided into three groups: those which are completely mixed and are vertically homogeneous; those which are partially vertically mixed and which are moderately stratified; and those with a salt wedge, which are highly stratified.

River, Reservoirs, Lakes and Wetland

Surface water is that water which occurs permanently or intermittently on the land sur-face in the form of different water bodies: rivers, streams and temporary watercourses, reservoirs, lakes, swamps, mires, glaciers, and snow cover. The process of transport-ing water of atmospheric origin over the land surface under the influence of gravity is called surface runoff. It is measured as a water discharge ($m^3 \cdot s^{-1}$), volume of water run-off (km^3), specific water discharge ($l \cdot s^{-1}km^{-2}$) or a layer of water run-off (mm) per year or some other period of time.

A river is a watercourse flowing in a self-developed bed augmented by surface and groundwater. With all its tributaries it forms a river system whose character and devel-opment is related to climate, relief, geologic structure, and the dimensions of the ba-sin. Rivers can be subdivided into mountain rivers – usually flowing rapidly in narrow valleys – and plain ones, which flow more slowly in wide terracing valleys. The water regime depends mainly on the character of river augmentation and the climate conditions in the region. Total annual river runoff into the World Ocean is about 47,000 km³.

A lake is a natural reservoir filled with water within a lake basin not directly linked with the sea. Basins are subdivided according to their origin into tectonic, glacial, fluvial, coastal, sinkhole (in karst and thermokarst), volcanic, and dammed (artificial reservoirs and ponds). Depending on the nature of the lake bed, three main lake types can be distinguished:

- Dammed lakes: fluvial, valley, and coastal (including reservoirs).

- Hollow lakes: moraine, karst, thermokarst, deflation, volcanic, tectonic.

- Lakes of mixed origin.

There are also other classifications. According to the water regime, lakes may be subdivided into lakes with an outlet (exorheic) and lakes without an outlet (endorheic). Exorheic lakes are those from which rivers flow, for example Lake Ontario (Canada/USA), and the Baikal, Onega, and Ladoga lakes in Russia. A particular case of such lakes is comprised of flowing lakes (drainage lakes). Rivers run into such lakes, and then flow out from them. The best-known flowing lakes, are Lakes Boden and Geneva, which allow the passage of the Rhine and Rhone rivers. From endorheic lakes no rivers flow. All the water that enters such a lake is taken up by evaporation or infiltration, or used for economic purposes. The largest endorheic lakes are the Caspian and Aral Seas, Issyk-Kul Lake, Lake Eyrie and Salt Lake.

A water reservoir is an artificial water basin, usually formed in a river valley by water supply lines that regulate its use for purposes of the natural economy. According to the form of the basin, reservoirs may be subdivided into fluvial, lacustrine, and mixed. Water reservoirs can be split into permanent or temporary (a day, a week, a season, or a year).

Vertical temperature stratification, caused by the fact that water reaches its maximum density at a temperature of 4° C, is characteristic of weakly flowing and deep lakes and reservoirs. In the warm period of the year, a uniform layer (the epilimnion) develops near the surface of lakes and reservoirs, and the oxygen regime is favorable to development of aquatic life. There is a layer below the epilimnion where the water temperature falls abruptly, and below which the density increases with depth. This layer is the thermocline. Deeper layers, in the hypolimnion, interact poorly with atmosphere, and the dissolved oxygen content decreases considerably, so that anaerobic processes occur with the liberation of hydrogen sulfide and methane. This phenomenon is particularly characteristic of many water reservoirs in arid and tropical zones.

A wetland is an area of land characterized by constant or excess moistening, favoring hydrophilic vegetation and the development of specific soil processes. Soil formation alternates with short periods of peat formation, followed by its washout. In some cases siltation with organic-mineral muds occurs, and sometimes it is only gleying.

Bogs (marshes and wetland) are formed through the excess moistening of area or eutrophication of a reservoir. The former is the most common. Bogs occur as a result of flooding of territories. Flooding may be observed in the coastal zone of rivers, lakes, reservoirs, and seas as a result of long-term rise in water levels. Under-flooding is also a consequence of this process. Bogs can also form when there is a change in the ratio between deposition and evaporation of atmospheric precipitation, and a natural or anthropogenic increase of subsoil waters. Eutrophication of reservoirs and their transformation into bogs is observed in zones of temperate and warm climates. It reflects a natural stage in the evolution of a reservoir.

The general area of bogs on the Earth equals 2.7×10^6 km² which amounts to about 2 percent of land area. They hold 0.03 percent of the freshwater on the planet. Eurasia and North America contain the greatest areas of bog.

Peat bogs are distributed in tundra, and the forest and forest-steppe zone. They differ among themselves in a series of attributes. Low-lying bogs have a concave or flat surface. They are formed on the shores of rivers, lakes and reservoirs, and in river mouths. Low-lying bogs are directly connected to the rivers and underground waters that provide them with mineral substances and good conditions of nutrient supply for aquatic communities. High-lying bogs generally have a convex surface, typically with a thick layer of peat. Such bogs are formed on drainage divides (watersheds), and also as a result of the growth of low-lying bogs. On watersheds, the water balance of bogs is basically determined by atmospheric precipitation and evaporation. Receipt of mineral substances is low, and this has a fundamental effect on the aquatic ecosystem. The intermediate type of peat bogs has a weakly convex surface and relatively better conditions of nutrient supply in comparison with high-lying bogs.

Depending on conditions of rainfall, water-temperature, and inflow, mires can be subdivided into oligotrophic (nutrient-poor) bogs, with low pH and low levels of nutrients available for vegetation, and eutrophic fens and marshes. Oligotrophic bogs receive most of their water from precipitation, and so tend to be located near river basin watersheds. Rainfall is high, so nutrients are constantly removed. Eutrophic mires, on the other hand, tend to be lowland in character, and the vegetation is not severely limited by lack of nutrients. They receive mineral nutrition from surface drainage and groundwater. They tend to be located in depressions, so minerals and nutrients can be provided from higher ground. There is a whole spectrum of mires of intermediate type where minerals are available to at least some parts of the area from groundwater inflow, while other parts are oligotrophic. Bogs can be further subdivided according to the dominant vegetation (scrub, dwarf shrubs, grasses, bogmosses, etc.) by micro-relief (hilly, flat, or domed), and macro-relief (valley, flood plain, slope, watershed), and other aspects.

Groundwater

This term refers to water in the Earth's crust in all physical states, in the sedimentary rock layers and massive-crystallized rock fractures. There are many groundwater classifications relying on different types of groundwater infiltration and distribution, lithological composition, geological age, and on differences in hydrodynamics, temperature, and chemical composition.

According to the conditions of their occurrence in rocks, the following types of groundwater are distinguished: free gravitational, in a solid state vaporous, physically bound, chemically bound, and water in a supercritical state. Free gravitational water, filling pores, voids, and fractures in rocks is the most widespread. This water percolates through rocks under a pressure drop. According to their ability to allow water to infiltrate, rocks can be classified as, on the one hand, water-bearing or water-permeable (loams, sandy clays) and, on the other, water-impermeable, or confining strata (clays, compact sandstones, and non-fractured rocks). Here, different types of water-bearing rocks can be distinguished: porous, fractured, fractured–porous, and karsts.

Water-bearing and poorly permeable or confining rocks usually occur as inter-bedded layers or zones of different thickness. Thus there is often the possibility of typifying a rock mass by identifying the aquifers and poorly permeable or confining beds within it.

Gravitational groundwater is mainly recharged by infiltrating rainfall, water vapor condensation (mainly in mountainous regions), and river runoff. This groundwater is discharged into rivers, gullies and ravines, seas and oceans, and also taken up by transpiration of plants and evaporation from the surface.

Groundwater is also subdivided according to its water exchange capacity. A zone of active water exchange is usually present near the surface, and is mainly characterized by fresh bicarbonate- and calcium-rich water. Then the occurs a zone of reduced water exchange with brackish, mainly calcium- and sodium-sulphate water, and, below this, a zone of very reduced water exchange, mainly with sodium- and calcium-chloride saline water. This zonality is disturbed in some areas due to peculiarities of hydrogeological conditions (e.g. the availability of salt-bearing or gypsum rocks, tectonic disturbances, etc. in the profile), causing the occurrence of so-called azonal groundwater.

The chemical composition of groundwater includes a mixture of many chemical elements, in the form of different ion types, neutral molecules, organic-mineral complexes, colloids, and isotopes. A complex of climatic, physical-geographical, soil-vegetation, structural-geologic and hydrogeological factors, produces groundwater from both infiltration and condensation. According to its general mineralization, groundwater is usually subdivided into fresh (with up 1 g/l mineralization), saline (from 1 to 25 g/l mineralization) and brine, with mineralization exceeding 25 g/l (in individual cases up to some hundred g/l).

When typifying groundwater according to its use, mineral water can be singled out on account of the concentrations of dissolved mineral salts, gases, and organic matters that allow its use for balneological treatment. The temperature of some groundwater allows it to be used as a thermal power source. Thermal water of a specific chemical composition can be also used for medical treatment.

Ground and surface water are often closely interrelated. Their interconnection is characterized by two opposite processes: augmentation of surface streams and reservoirs and groundwater recharge from surface water. A combination of these two processes is possible within one river basin both in time and space but the processes are considerably affected by groundwater exploitation. Processes of groundwater interrelations, caused primarily by geologic-hydrogeologic conditions of the river basins in the coastal zones, should be considered when assessing total water resources and the water balance of individual regions, in addition to groundwater withdrawal for different purposes.

Soil Water

The term "soil water" usually refers to the water localized in soil pore space (i.e. in the surficial part of the land) in the form of liquid moisture (both closely attached to soil skeletal particles and water freely able to move through the soil profile), a solid component in the form of ice in the soil pore space, and gaseous water in the form of soil air.

The proportion of soil water in the total volume of inland water is very small (0.06 percent). However its geo- and biophysical function on the Earth is no less important than that of groundwater. This function is connected with the boundary character of soil and soil water. Solar radiation energy reaching the Earth and driving the global circulation system is transformed into other forms of energy in a very thin planetary layer, on the boundary between atmosphere and lithosphere. This is where all the four components of the biosphere interact: the atmosphere, the upper part of the lithosphere, the hydrosphere, and terrestrial living mater.

Soil water occurs in the zone where solar radiation is transformed and assimilated by the biosphere. It thus has an important role, clearly accounting for three of the most important inter-related aspects and determining their place in the Earth hydrosphere. First, soil water is the most active link in mutual exchange of land water. In the terrestrial hydrological cycle, the soil layer serves as a specific "water separator," controlling the partition distribution of incoming water (precipitation) into its three outgoing components: surface runoff, subsurface runoff, and evapotranspiration. Second, soil water is an important element of the Earth's climatic system. Third, soil water is the most important factor controlling the presence and growth of vegetation cover; in other words, it is a primary link in the trophic system of terrestrial ecosystems. The dynamics of soil water stores has a major influence on vegetation cover.

As an important component of the hydrogeological cycle, evapotranspiration deserves special mention. Global evapotranspiration (of which 80–90 percent is transpiration) represents about two-thirds of the overall precipitation on the land surface. It seems that a great water volume precipitates on the land surface to no purpose. Indeed, a plant needs no more than 1–3 percent of all the water it uses in transpiration. This small amount is required for the production of plant material through photosynthesis, and to maintain the required concentrations and osmotic pressures. The remaining 97–99 percent is returned to the atmosphere. "It can be compared to using Niagara Falls to fill a bath." However, analysis of terrestrial vegetation in relation to the theory of dissipative structures, and consideration of their role in production of entropy, which is permanently produced by plants, demonstrates that during the evolution of life on Earth, transpiration may be the only mechanism that can do this without causing temperature increases which would damage plant tissues. Thus, transpiration in terrestrial vegetation (the expendable component of the soil water balance) is an important part of global biogeochemical cycling.

Glaciers, Icebergs and Ground Ice

Ice is the most abundant "'mineral'" on the Earth. The total mass of ice enclosed in glaciers, icebergs, ground ice, snow cover, and the atmosphere is 2.423 x1022 tons. Ice covers more than 16.3×10^6 km², or 11 percent of the Earth surface. The total ice volume of modern glaciers ranges from 26.8×10^6 km³ to 30.3×10^6 km³. If the ice layer covered the Earth uniformly, its thickness would be approximately 55–60 meters.

A glacier is a moving natural accumulation of ice on the land, under a negative balance of the solid phase of water. Glaciers are confined to those places on the Earth surface where solid precipitation exceeds evaporation. Most glaciers consist of an alimentation zone, where the snow is accumulated and transformed into firn and ice, and an ablation zone, where ice is lost from the glacier by melting and evaporation. There are two main types of glaciers: mountainous (flowing down) and covering (spreading). Net-shaped and piedmont types of glaciation are transient between mountainous and covering glaciers.

Mountain glaciers occupy mainly negative elements of the relief, forming cirque (corrie), valley, and other types of glaciers; ice moves slowly downslope under gravitational force. In many dry areas of the Earth, mountain glaciers supply a considerable part of the water used for irrigation. Glaciers are only found in the vicinity of the snow line. Above the snow line the accumulation of solid precipitation is greater than the combined thawing, evaporation, and run off. The level of the snowline oscillates widely, depending on the moisture and heat balance and local climatic conditions Its altitude may vary from sea level in the Antarctic Continent to 6,000– 6,500 m above sea level on the Tibetan Plateau.

In Europe modern glaciers are concentrated in Scandinavia, the Alps, the Caucasus, and the Urals. There are small glaciers in the Khibiny and Perinea Mountains. There are 9,529 glaciers in Europe, with a total area of 7,395 km^2 . In Asia, ice covers high mountains areas of the Tien Shan, Pamirs, Karakorum, and Himalayas, where many very large dendritic and complex mountain valley glaciers descend well below the snow line. The largest centers of modern glaciation in Northern America are located in the Northern America Cordillera. The Alaska Ridge also has large centers of glaciation, which aliment from the Pacific.

Cover glaciers spread over many million square kilometers, blanketing even mountains with a domed surface, moving slowly from the center to the periphery in a radiating pattern. More than 96.6 percent of the area of cover glaciers and 90 percent of the volume of ice are concentrated in the Greenland and Antarctic ice sheets. The Antarctic Continent covers 14·106 km^2; the mean diameter of the Antarctic ice sheet is 4,000 km, with a minimum of 2,900 km, and a maximum of 5,500 km. Most glaciers of Eurasia and the Canadian Arctic are classified as ice caps by their morphology, and as diffluent glaciers by their movement features. The largest Eurasian ice caps are located in Iceland, Franz Josef Land, Spitsbergen, Novaya Zemlya Island, Severnaya Zemlya Island, Bennett Island, Henriette Island, Jeannette Island, Victoria Island, Ushakov Island, and Schmidt Island. The largest ice caps of the Canadian Arctic Archipelago are those of Baffin Land, Ellesmere Island, Devon Island, Axel-Heiberg, Melville Island, and Meighen Island. To the north of Ellesmere Island, there is a small shelf glacier, on Ward Hunt Island, that produces table icebergs. Shelf ice, partially resting on the sea bottom, is a continuation of terrestrial ice and is mainly found in the Antarctic.

Icebergs are masses of continental ice, which have separated from a glacier or glacial barrier and float in the polar and adjoining seas and oceans. Continental ice from Antarctica, Greenland, and the glaciers of Severnaya Zemlya is the major center of iceberg formation. Due to the similarity of the densities of ice and seawater, the greater part of an iceberg (80–85 percent of the total height) is under the water. Icebergs are moved by currents and are gradually destroyed by thawing and weathering. There are the following types of iceberg: tabular, domelike, pyramidal, and destroyed. Tabular icebergs have a flat surface because they usually break off shelf glaciers. They may be several kilometers in length and width. They are often found around the Antarctic Continent, but sometimes they form near Greenland and around the ice caps of Arctic Islands. Domelike icebergs mainly come from diffluent glaciers or from ice bluffs; their height is about 70–100 m. Pyramidal icebergs are pyramidal in shape, and destroyed icebergs have an irregular shape, often with several summits. Separation of icebergs sometimes happens as a result of tides. A large iceberg can traverse a great distance, more than 4,000–6,000 km from its place of origin.

Snow cover is a layer of snow, produced by snowfalls and lying on the land or ice surface. According to its appearance and conditions of formation, snow can be subdivided into newly fallen, compacted (stable) and old (firnified, neve). The density of snow varies from 0.01 g/cm^3 for newly fallen to 0.70 g/cm^3 for highly wet and then frozen forms. Globally, snow covers an area from 115 to 126 x 10^6 km^2 (about a third of this area is sea ice). Snow reflects solar radiation, and protects soil from overcooling and winter crops from freezing.

Solid atmospheric precipitation, accumulated on the land surface, changes considerably over time. During snow accumulation there is freezing and melting of snowflakes, their compaction, and structural change. As a result, snow layers are transformed into a porous, white-gray mass, called firm. In the process of being compacted, firn (which has a density of 0.3-0.5 g/cm³) is first transformed into white ice with a density of about 0.85 g/cm³, and then into transparent ice with a density of 0.91 g/cm³. Pieces of ice can grow together into a uniform mass (regelation). Another important ice property is its plasticity, which allows it to move under gravitation. Ice is the most widespread solid material on the Earth, forming glaciers, icebergs, and marine and ground ice.

Ground ice is a general term used to refer to all types of ice formed in freezing and frozen ground. Ground ice occurs in the pores, cavities, voids, or other openings in soil or rock and includes massive ice. Buried glacier, lake, river, and snow bank ice are all categorized together as a single type of ground ice. Traditionally permafrost is defined on the basis of temperature: soil or rock that remains at or below 0° C for at least at two consecutive years.

The ice content in permafrost is probably the most important feature relevant to human life in the northern regions of Siberia, Scandinavia, Alaska, Canada, and sub-polar regions of the Southern Hemishpere. Ice in perennially frozen ground exists in various sizes and shapes, with definite distribution characteristics grouped into five main types: pore ice, segregated or taber ice, foliated ice or ice wedges, pingo ice, and buried ice. Around 10 percent by volume of the upper 3–5 m of the permafrost of the Siberian, Alaskan, and Canadian coastal plain is composed of ice wedges. Taber ice is the most extensive type, in places representing 75 percent of the ground ice by volume.

Uses of Water Resources

1. Transportation: Water in the rivers, lakes, and oceans are often used as a means of transport for conveying people and goods from one place to another for example Atlantic Ocean, river Niger and river Benue.

2. Hydroelectric Power: These various sources of water presents on the surface of the earth are demanded for the production of electricity for example the Kainji Dam, Akomsombo dam (river Volta in Ghana).

3. Mineral Deposit: Water bodies serve a source of mineral deposits for example, salt, Placer Gold, Tin, crude oil, Titanium, and Diamonds, Limestone and Gypsum are extracted from large water bodies.

4. Source of Food Man: Edible foods such as seaweeds and microalgae are widely eaten asseafoodby different humans around the world. However, shellfish like

Squid, crab, octopus, oyster, shrimp, and lobster are harvested from saltwater environments.

5. Employment: The various water bodies such as lakes, rivers and oceans provide employment opportunities for people in various locations of the world. Ship builders, sailors, mariners, canoe carvers and fishermen are examples of such people.

6. Recreation and Tourism: Various water bodies also provide excellent facilities for swimming fishing and picnicking, boating skiing. Even activities such as golf, where there may not be any standing water, require plenty of water to make the grass on the course green. A larger water body like oceans provides beautiful scenery which attracts tourists.

7. Domestic Use: Globally, household or personal water use is estimated to account for 15% of worldwide water use; since water is one of the most important necessities of life, it is highly needed in our homes for various uses. Domestic water use is water used for indoor and outdoor household purpose for instance for drinking, preparing food, bathing, washing of clothes and dishes, brushing of teeth, watering of yard and garden.

8. Industrial use: A large quantity of water is required by large industries like hydroelectric dams, thermoelectric power plants for cooling and generation of power. Large industries like oreandoilrefineries use water in chemical processes and manufacturing plants also use water as a solvent. Other example include various breweries, Iron smelting companies which use water for washing and processing of their raw materials.

9. Agricultural use: Some of the worlds farmers still farm without irrigation by choosing crops that match the amount of rain that falls in their area. However, some years are wet and others are extremely dry, in this case, farmer use water for irrigation to produce crops all year round.

10. Fishing: Fishing is a major activity which is common among the riverine people through which they earn their living. This activity is usually carried out all year round as to supply fish to various locations locally and internationally.

11. Environmental use: Environmental water may include water stored in impoundments and released for environmental purposes, but more often is water retained in waterways through regulatory limits of abstraction. Environmental water usage includes creating wildlife habitat, watering of natural or artificial wetlands, artificial lakes intended to create wildlife habitat for fish, water birds, and water releases from reservoirs timed to help fish spawn, or to restore more natural flow regimes.

Hydropower

Figure: Three Gorges Dam (left), Gezhouba Dam (right).

The energy of falling water and running water can be utilized to provide *water power* or *hydropower*—the form of renewable energy derived from the gravitational force of falling or flowing water harnessed for useful purposes. Since water is about 800 times denser than air, even a slow flowing stream of water, or moderate sea swell, can yield considerable amounts of energy. Since ancient times, hydropower has been used for irrigation and the operation of various mechanical devices, such as watermills, sawmills, textile mills, dock cranes, domestic lifts, and power houses.

Since the early 20th century, the term hydropower has been used almost exclusively in conjunction with the modern development of hydroelectric power, which allowed use of distant energy sources. *Hydroelectricity* is the term referring to electricity generated by hydropower; the production of electrical power through the use of the gravitational force of falling or flowing water.

There are many forms of water power:

- Hydroelectric energy from large-scale hydroelectric dams. The largest example of a large-scale hydroelectric dam is the Three Gorges Dam in China and a smaller example is the Akosombo Dam in Ghana.

- Micro hydro systems are hydroelectric power installations that typically produce up to 100 kW of power. They are often used in water rich areas as a remote-area power supply (RAPS).

- Run-of-the-river hydroelectricity systems derive kinetic energy from rivers and oceans without the creation of a large reservoir.

- Trompe is another method used to transmit energy. A trompe is a water-powered gas compressor, commonly used before the advent of the electric-powered compressor, which is somewhat like an airlift pump working in reverse. A trompe produces compressed air from falling water. Compressed air could then be piped to power other machinery at a distance from the waterfall.

Hydroelectricity today is the most widely used form of renewable energy (unless all biomass categories, such as wood and biofuels, are lumped together), accounting for 16 percent of global electricity generation—3,427 terawatt-hours of electricity production in 2010. Hydropower is produced in at least 150 countries, with five countries (China, Brazil, United States, Canada, and Russia) accounting for about 52 percent of the world's installed hydropower capacity in 2010. The Asia-Pacific region generated 32 percent of global hydropower in 2010, with China being the largest hydropower producer, producing 721 terawatt-hours in 2010 and having the highest installed hydropower capacity, with 213 gigawatts (GW) at the end of 2010. The Three Gorges Dam, spanning China's Yangtze River, is the world's largest hydroelectric power station in terms of installed capacity, followed in second place by the Itaipu Dam in Brazil/Paraguay. The Three Gorges Dam is operated jointly with the much smaller Gezhouba Dam.

The cost of hydroelectricity is relatively low, making it a competitive source of renewable electricity. The average cost of electricity from a hydro plant larger than 10 megawatts is 3 to 5 U.S. cents per kilowatt-hour. Hydro is also a flexible source of electricity since plants can be ramped up and down very quickly to adapt to changing energy demands. Once a hydroelectric complex is constructed, the project produces no direct waste, and has a considerably lower output level of the greenhouse gas carbon dioxide (CO_2) than fossil fuel powered energy plants. However, damming interrupts the flow of rivers and can harm local ecosystems, and building large dams and reservoirs often involves displacing people and wildlife. Given such concerns, in some nations building new dams on major rivers to capture hydroelectric energy meets a lot of resistance and further expansion of hydropower in the United States is unlikely. On the other hand, China's Three Gorges Dam became fully functional in just 2012.

Types of Hydropower

The main difference between these 4 types of hydropower is the method at which the water is stored and used to pass the turbine. The differences are:

1. Storage Hydropower

This is the most well-known type of hydropower. It uses a dam to create a massive water reservoir downstream of a river. The generation of electricity can easily be controlled by adjusting the water flow (volumetric flow rate) that passes through a turbine from the reservoir. Water flow must be higher during times of high electrical use (peak load) and lower during times when the minimum electrical output is required (base load). Water is also released to match a specific water level for the reservoir.

2. Run of river hydropower

Normally, run of river does not need a storage reservoir because it uses the natural current of the river to rotate a turbine. A separate canal or penstock is constructed in the river in order to divert some of the water flow to be used to generate electricity. As a result, it generates a continuous supply of electricity. During times of low electrical use, water flow is limited in the canals.

3. Pumped Storage Hydropower

Two reservoirs are used to reuse the same supply of water. One reservoir is placed at a higher elevation and another at a lower elevation. During peak hours of electrical use, the higher elevated reservoir is opened to allow water to flow through a turbine towards the lower reservoir. During times of low use such as dawn, water is pumped back to the higher elevated reservoir from the lower reservoir for future use.

These 3 types of hydropower can also be used in different combinations. For example, if the water level of a storage facility is lower than the specified level and the natural flow of the river does not fill it back up, water can be pumped from other water sources.

Daytime: Water flows downhill through turbines, producing electricity

Nightime: Water pumped uphill to reservoir for tomorrow's use

4. Offshore hydropower

Offshore hydropower is a less established but growing group of technologies that use tidal currents or the power of waves to generate electricity from seawater.

Sizes of Hydroelectric Power Plants

Facilities range in size from large power plants that supply many consumers with electricity to small and micro plants that individuals operate for their own energy needs or to sell power to utilities.

Large Hydropower

Although definitions vary, DOE defines large hydropower as facilities that have a capacity of more than 30 megawatts (MW).

Small Hydropower

Although definitions vary, DOE defines small hydropower as projects that generate 10 MW or less of power.

Micro Hydropower

A micro hydropower plant has a capacity of up to 100 kilowatts. A small or micro-hydroelectric power system can produce enough electricity for a home, farm, ranch, or village.

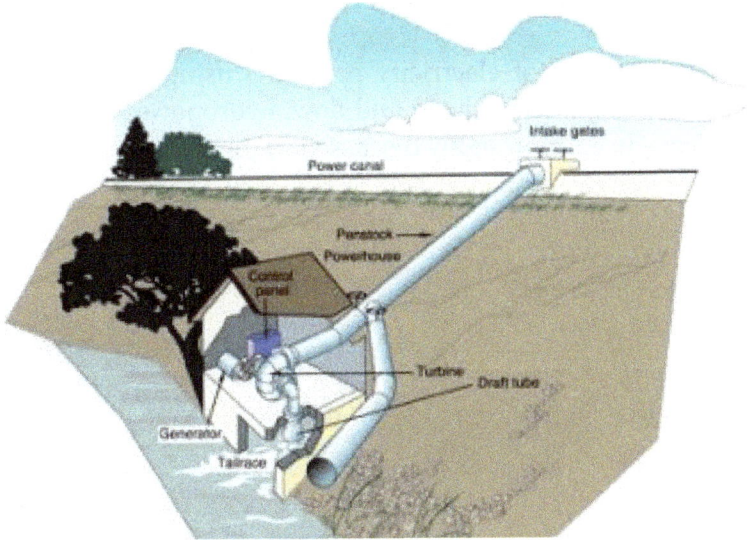

References

- What-are-water-resources: wisegeek.com, Retrieved 12 April 2018

- 3-types-hydropower: greenfuture.info, Retrieved 30 March 2018

- Water-resources-meaning-sources-of-water-importance-uses-of-water-211: jotscroll.com, Retrieved 19 May 2018

- Hydropower, Renewable-energy: newworldencyclopedia.org, Retrieved 12 June 2018

- Types-hydropower-plants, water: energy.gov, Retrieved 10 July 2018

2

Hydroelectricity

Hydroelectricity is the form of electricity produced using hydropower. The diverse aspects of hydroelectricity such as head and flow, hydropower potential, tidal energy, wave power, etc. have been thoroughly discussed in this chapter.

The Nagarjuna dam & hydroelectric plant, India

Hydroelectricity is electricity produced by hydropower—that is, the energy of moving water. It is the world's leading form of renewable energy. In 2005, it accounted for over 63 percent of the total renewable energy. That same year, hydroelectricity supplied about 715,000 megawatts (or 19 percent) of the world's electricity (compared to 16 percent in 2003). Although large hydroelectric installations generate most of the world's hydroelectricity, small hydro schemes are particularly popular in China, which has over 50 percent of the world's small hydro capacity.

On the downside, hydroelectric projects can disrupt ecosystems by reducing access to salmon spawning grounds and damaging habitats of birds. These projects also lead to changes in the downstream river environment, by the scouring of river beds and loss of riverbanks. Large-scale hydroelectric dams, such as the Aswan Dam and the Three Gorges Dam, have created environmental problems both upstream and downstream.

World Renewable Energy 2005

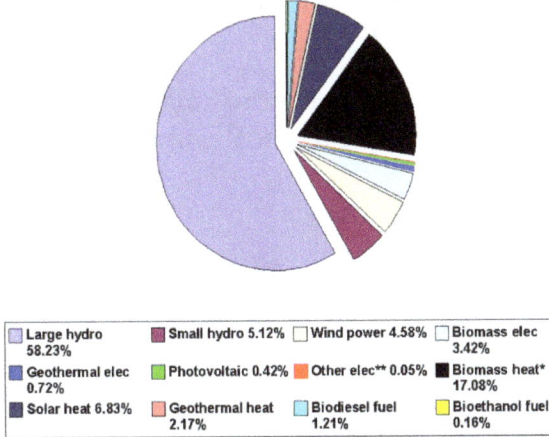

Large hydro 58.23%	Small hydro 5.12%	Wind power 4.58%	Biomass elec 3.42%
Geothermal elec 0.72%	Photovoltaic 0.42%	Other elec** 0.05%	Biomass heat* 17.08%
Solar heat 6.83%	Geothermal heat 2.17%	Biodiesel fuel 1.21%	Bioethanol fuel 0.16%

Hydroelectricity is the world's leading renewable energy source

Electricity Generation

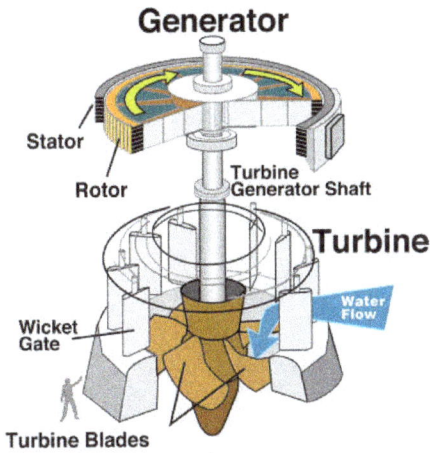

Hydraulic turbine and electrical generator

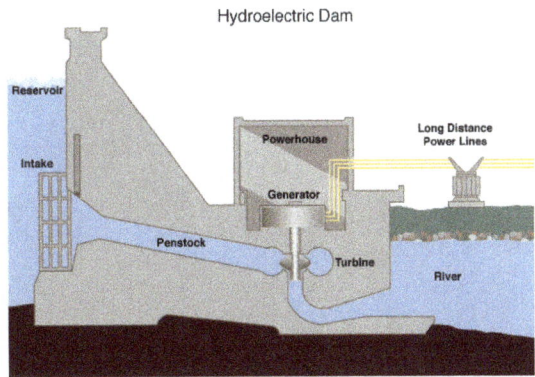

Hydroelectric dam in cross section

Most hydroelectric power comes from the potential energy of dammed water driving a water turbine and generator. In this case, the energy extracted from the water depends on the volume and on the difference in height between the source and the water's outflow. This height difference is called the head. The amount of potential energy in water is proportional to the head. To obtain very high head, water for a hydraulic turbine may be run through a large pipe called a penstock.

Pumped storage hydroelectricity produces electricity to supply high peak demands by moving water between reservoirs at different elevations. At times of low electrical demand, excess generation capacity is used to pump water into the higher reservoir. When there is higher demand, water is released back into the lower reservoir through a turbine. Pumped storage schemes currently provide the only commercially important means of grid energy storage and improve the daily load factor of the generation system.

Less common types of hydro schemes use water's kinetic energy or undammed sources such as run-of-the-river, waterwheels, and tidal power.

Industrial Hydroelectric Plants

Many hydroelectric projects supply public electricity networks, but some are created to serve specific industrial enterprises. Dedicated hydroelectric projects are often built to provide the substantial amounts of electricity needed for aluminium electrolytic plants, for example. In the Scottish Highlands there are examples at Kinlochleven and Lochaber, constructed during the early years of the twentieth century. In Suriname, the "van Blommestein" lake, dam and power station were constructed to provide electricity for the Alcoa aluminium industry. As of 2007 the Kárahnjúkar Hydropower Project in Iceland remains controversial.

Advantages

Economics

The upper reservoir and dam of the festiniog pumped storage scheme. 360 megawatts of electricity can be generated within 60 seconds of the need arising

The main advantage of hydroelectricity is that it is nearly independent of increases in the cost of fossil fuels such as oil, natural gas, or coal. Fuel is not required, and so it need not be imported. Hydroelectric plants tend to have longer economic lives than fuel-fired generation, with some plants now in service having been built 50 to 100 years ago. Operating labor cost is usually low since plants are automated and have few personnel on site during normal operation.

Where a dam serves multiple purposes, a hydroelectric plant may be added with relatively low construction cost, providing a useful revenue stream to offset the costs of dam operation. It has been calculated that the sale of electricity from the Three Gorges Dam will cover the construction costs after 5 to 7 years of full generation.

Related Activities

Reservoirs created by hydroelectric schemes often provide facilities for water sports, and become tourist attractions in themselves. In some countries, farming fish in the reservoirs is common. Multi-use dams installed for irrigation can support the fish farm with relatively constant water supply. Large hydro dams can control floods, which would otherwise affect people living downstream of the project. When dams create large reservoirs and eliminate rapids, boats may be used to improve transportation.

Greenhouse Gas Emissions

Since no fossil fuel is consumed, emission of carbon dioxide (a greenhouse gas) from burning fuel is eliminated. While some carbon dioxide is produced during manufacture and construction of the project, this is a tiny fraction of the operating emissions of equivalent fossil-fuel electricity generation.

Disadvantages

Environmental Damage

Recreational users must exercise extreme care when near hydroelectric dams, power plant intakes and spillways

Hydroelectric projects can be disruptive to surrounding aquatic ecosystems. For instance, studies have shown that dams along the Atlantic and Pacific coasts of North America have reduced salmon populations by preventing access to spawning grounds upstream, even though most dams in salmon habitat have fish ladders installed. Salmon spawn are also harmed on their migration to sea when they must pass through turbines. This has led to some areas transporting smolt downstream by barge during parts of the year. Turbine and power-plant designs that are easier on aquatic life are an active area of research.

Generation of hydroelectric power changes the downstream river environment. Water exiting a turbine usually contains very little suspended sediment, which can lead to scouring of river beds and loss of riverbanks. Since turbines are often opened intermittently, rapid or even daily fluctuations in river flow are observed. For example, in the Grand Canyon, the daily cyclic flow variation caused by Glen Canyon Dam was found to be contributing to erosion of sand bars. Dissolved oxygen content of the water may change from pre-construction conditions. Water exiting from turbines is typically much colder than the pre-dam water, which can change aquatic faunal populations, including endangered species. Some hydroelectric projects also utilize canals, typically to divert a river at a shallower gradient to increase the head of the scheme. In some cases, the entire river may be diverted, leaving a dry riverbed. Examples include the Tekapo and Pukaki Rivers.

Large-scale hydroelectric dams, such as the Aswan Dam and the Three Gorges Dam, have created environmental problems both upstream and downstream.

A further concern is the impact of major schemes on birds. Since damming and redirecting the waters of the Platte River in Nebraska for agricultural and energy use, many native and migratory birds such as the Piping Plover and Sandhill Crane have become increasingly endangered.

Greenhouse Gas Emissions

The reservoirs of hydroelectric power plants in tropical regions may produce substantial amounts of methane and carbon dioxide. This is due to plant material in flooded areas decaying in an anaerobic environment, and forming methane, a very potent greenhouse gas. According to the World Commission on Dams report, where the reservoir is large compared to the generating capacity (less than 100 watts per square meter of surface area) and no clearing of the forests in the area was undertaken prior to impoundment of the reservoir, greenhouse gas emissions from the reservoir may be higher than those of a conventional oil-fired thermal generation plant.

In boreal reservoirs of Canada and Northern Europe, however, greenhouse gas emissions are typically only 2 to 8 percent of any kind of conventional thermal generation. A new class of underwater logging operation that targets drowned forests can mitigate the effect of forest decay.

Population Relocation

Another disadvantage of hydroelectric dams is the need to relocate the people living where the reservoirs are planned. In many cases, no amount of compensation can replace ancestral and cultural attachments to places that have spiritual value to the displaced population. Additionally, historically and culturally important sites can be flooded and lost. Such problems have arisen at the Three Gorges Dam project in China, the Clyde Dam in New Zealand, and the Ilısu Dam in Southeastern Turkey.

Dam Failures

Failures of large dams, while rare, are potentially serious—the Banqiao Dam failure in China resulted in the deaths of 171,000 people and left millions homeless, more than some estimates of the death toll from the Chernobyl disaster. Dams may be subject to enemy bombardment during wartime, sabotage, and terrorism. Smaller dams and micro hydro facilities are less vulnerable to these threats.

The creation of a dam in a geologically inappropriate location may cause disasters like the one of the Vajont Dam in Italy, where almost 2000 people died, in 1963.

Comparison with other Methods of Power Generation

Hydroelectricity eliminates the flue gas emissions from fossil fuel combustion, including pollutants such as sulfur dioxide, nitric oxide, carbon monoxide, dust, and mercury in the coal.

Compared to the nuclear power plant, hydroelectricity generates no nuclear waste, nor nuclear leaks. Unlike uranium, hydroelectricity is also a renewable energy source.

Compared to wind farms, hydroelectricity power plants have a more predictable load factor. If the project has a storage reservoir, it can be dispatched to generate power when needed. Hydroelectric plants can be easily regulated to follow variations in power demand.

Unlike fossil-fueled combustion turbines, construction of a hydroelectric plant requires a long lead-time for site studies, hydrological studies, and environmental impact assessment. Hydrological data up to 50 years or more is usually required to determine the best sites and operating regimes for a large hydroelectric plant. Unlike plants operated by fuel, such as fossil or nuclear energy, the number of sites that can be economically developed for hydroelectric production is limited; in many areas the most cost effective sites have already been exploited. New hydro sites tend to be far from population centers and require extensive transmission lines. Hydroelectric generation depends on rainfall in the watershed, and may be significantly reduced in years of low rainfall or snowmelt. Utilities that primarily use hydroelectric power may spend additional capital to build extra capacity to ensure sufficient power is available in low water years.

Principles of Hydropower Generation

Hydroelectric power comes from water at work, water in motion. It can be seen as a form of solar energy, as the sun powers the hydrologic cycle which gives the earth its water. In the hydrologic cycle, atmospheric water reaches the earth's surface as precipitation. Some of this water evaporates, but much of it either percolates into the soil or becomes surface runoff. Water from rain and melting snow eventually reaches ponds, lakes, reservoirs, or oceans where evaporation is constantly occurring.

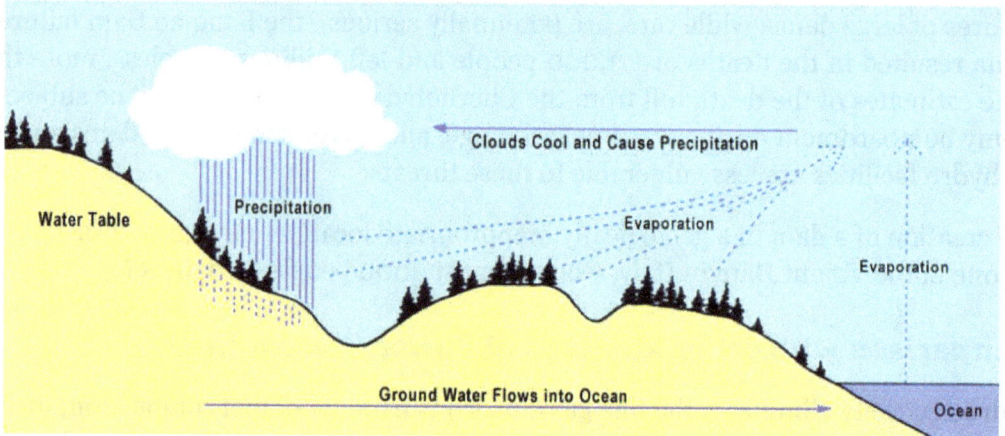

Moisture percolating into the soil may become ground water (subsurface water), some of which also enters water bodies through springs or underground streams. Ground water may move upward through soil during dry periods and may return to the atmosphere by evaporation.

Water vapor passes into the atmosphere by evaporation then circulates, condenses into clouds, and some returns to earth as precipitation. Thus, the water cycle is complete. Nature ensures that water is a renewable resource.

Generating Power

In nature, energy cannot be created or destroyed, but its form can change. In generating electricity, no new energy is created. Actually one form of energy is converted to another form.

To generate electricity, water must be in motion. This is kinetic (moving) energy. When flowing water turns blades in a turbine, the form is changed to mechanical (machine) energy. The turbine turns the generator rotor which then converts this mechanical energy into another energy form electricity. Since water is the initial source of energy, we call this hydroelectric power or hydropower for short.

At facilities called hydroelectric power plants, hydropower is generated. Some power plants are located on rivers, streams, and canals, but for a reliable water supply, dams are needed. Dams store water for later release for such purposes as irrigation, domestic and industrial use, and power generation. The reservoir acts much like a battery, storing water to be released as needed to generate power.

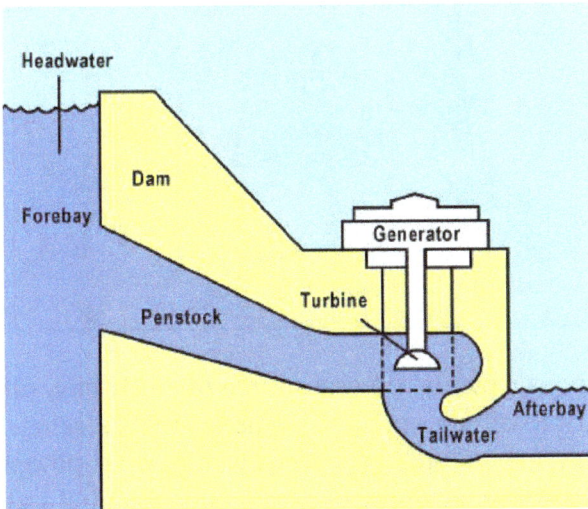

The dam creates a "head" or height from which water flows. A pipe (penstock) carries the water from the reservoir to the turbine. The fast-moving water pushes the turbine blades, something like a pinwheel in the wind. The waters force on the turbine blades turns the rotor, the moving part of the electric generator. When coils of wire on the rotor sweep past the generator's stationary coil (stator), electricity is produced.

This concept was discovered by Michael Faraday in 1831 when he found that electricity could be generated by rotating magnets within copper coils.

When the water has completed its task, it flows on unchanged to serve other needs.

Transmitting Power

Once the electricity is produced, it must be delivered to where it is needed our homes, schools, offices, factories, etc. Dams are often in remote locations and power must be transmitted over some distance to its users.

Vast networks of transmission lines and facilities are used to bring electricity to us in a form we can use. All the electricity made at a power plant comes first through transformers which raise the voltage so it can travel long distances through power lines. (Voltage is the pressure that forces an electric current through a wire.) At local substations, transformers reduce the voltage so electricity can be divided up and directed throughout an area.

Transformers on poles (or buried underground, in some neighborhoods) further reduce the electric power to the right voltage for appliances and use in the home. When electricity gets to our homes, we buy it by the kilowatt-hour, and a meter measures how much we use.

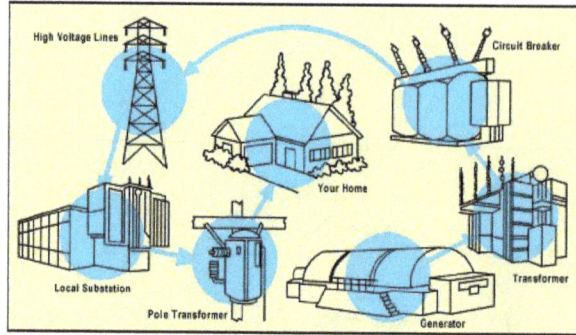

While hydroelectric power plants are one source of electricity, other sources include power plants that burn fossil fuels or split atoms to create steam which in turn is used to generate power. Gas turbine, solar, geothermal, and wind-powered systems are other sources. All these power plants may use the same system of transmission lines and stations in an area to bring power to you. By use of this a power grid," electricity can be interchanged among several utility systems to meet varying demands. So the electricity lighting your reading lamp now may be from a hydroelectric power plant, a wind generator, a nuclear facility, or a coal, gas, or oil-fired power plant or a combination of these.

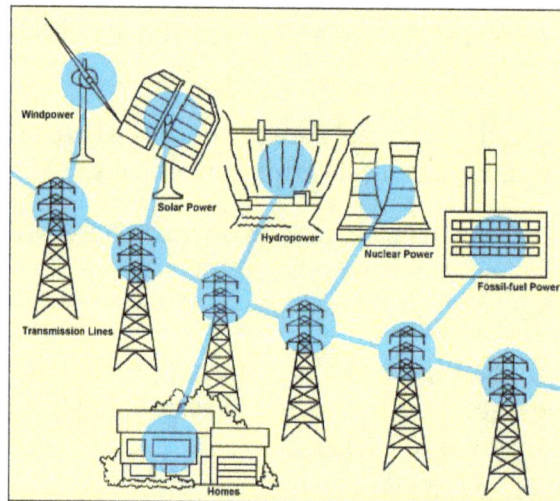

The area where you live and its energy resources are prime factors in determining what kind of power you use. For example, in Washington State hydroelectric power plants provided approximately 80 percent of the electrical power during 2002. In contrast, in Ohio during the same year, almost 87 percent of the electrical power came from coal-fired power plants due to the area's ample supply of coal.

Electrical utilities range from large systems serving broad regional areas to small power companies serving individual communities. Most electric utilities are investor-owned (private) power companies. Others are owned by towns, cities, and rural electric associations. Surplus power produced at facilities owned by the Federal Government is marketed to preference power customers (A customer given preference by law in the purchase of federally generated electrical energy which is generally an entity which is nonprofit and publicly financed.) by the Department of Energy through its power marketing administrations.

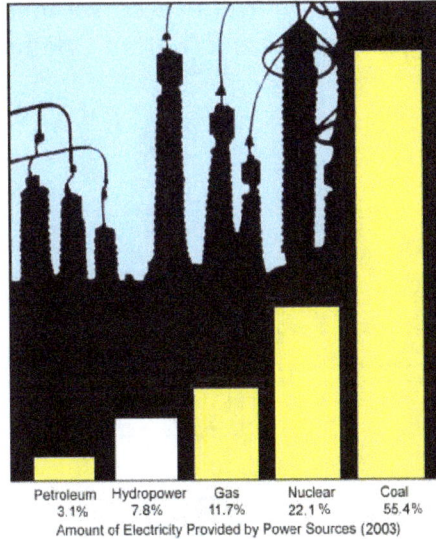

Petroleum Hydropower   Gas   Nuclear   Coal
   3.1%   7.8%   11.7%   22.1 %   55.4%
Amount of Electricity Provided by Power Sources (2003)

Computation of Power

Before a hydroelectric power site is developed, engineers compute how much power can be produced when the facility is complete. The actual output of energy at a dam is determined by the volume of water released (discharge) and the vertical distance the water falls (head). So, a given amount of water falling a given distance will produce a certain amount of energy. The head and the discharge at the power site and the desired rotational speed of the generator determine the type of turbine to be used.

The head produces a pressure (water pressure), and the greater the head, the greater the pressure to drive turbines. This pressure is measured in pounds of force (pounds per square inch). More head or faster flowing water means more power.

To find the theoretical horsepower (the measure of mechanical energy) from a specific site, this formula is used:

THP = (Q x H)/8.8

Where: THP = theoretical horsepower

Q = flow rate in cubic feet per second (cfs)

H = head in feet

8.8 = a constant

A more complicated formula is used to refine the calculations of this available power. The formula takes into account losses in the amount of head due to friction in the penstock and other variations due to the efficiency levels of mechanical devices used to harness the power.

To find how much electrical power we can expect, we must convert the mechanical measure (horsepower) into electrical terms (watts). One horsepower is equal to 746 watts (U.S. measure).

Turbines

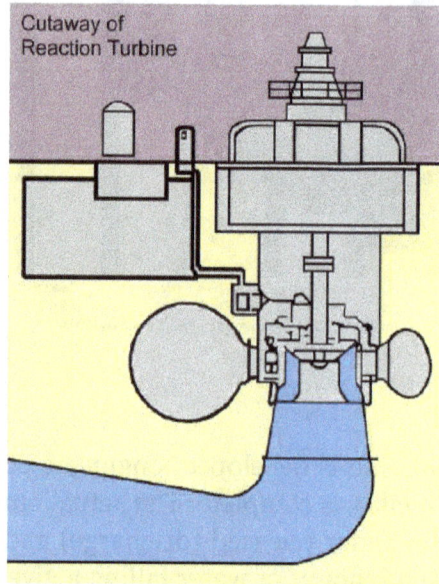

Cutaway of
Reaction Turbine

While there are only two basic types of turbines (impulse and reaction), there are many variations. The specific type of turbine to be used in a power plant is not selected until all operational studies and cost estimates are complete. The turbine selected depends largely on the site conditions.

A reaction turbine is a horizontal or vertical wheel that operates with the wheel completely submerged, a feature which reduces turbulence. In theory, the reaction turbine works like a rotating lawn sprinkler where water at a central point is under pressure and escapes from the ends of the blades, causing rotation. Reaction turbines are the type most widely used.

An impulse turbine is a horizontal or vertical wheel that uses the kinetic energy of water striking its buckets or blades to cause rotation. The wheel is covered by a housing and the buckets or blades are shaped so they turn the flow of water about 170 degrees inside the housing. After turning the blades or buckets, the water falls to the bottom of the wheel housing and flows out.

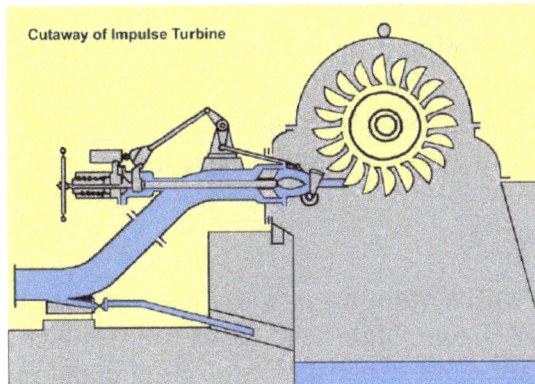

Cutaway of Impulse Turbine

Modern Concepts and Future Role

Hydropower does not discharge pollutants into the environment; however, it is not free from adverse environmental effects. Considerable efforts have been made to reduce environmental problems associated with hydropower operations, such as providing safe fish passage and improved water quality in the past decade at both Federal facilities and non-Federal facilities licensed by the Federal Energy Regulatory Commission.

Efforts to ensure the safety of dams and the use of newly available computer technologies to optimize operations have provided additional opportunities to improve the environment. Yet, many unanswered questions remain about how best to maintain the economic viability of hydropower in the face of increased demands to protect fish and other environmental resources.

Existing hydropower concepts and approaches include:

- Uprating existing power plants

- Developing small plants (low-head hydropower)

- Peaking with hydropower

- Pumped storage

- Tying hydropower to other forms of energy

Uprating

The uprating of existing hydroelectric generator and turbine units at power plants is one of the most immediate, cost-effective, and environmentally acceptable means of developing additional electric power. Since 1978, Reclamation has pursued an aggressive uprating program which has added more than 1,600,000 kW to Reclamation's capacity at an average cost of $69 per kilowatt. This compares to an average cost for providing new peaking capacity through oil-fired generators of more than $400 per kilowatt. Reclamation's uprating program has essentially provided the equivalent of another major hydroelectric facility of the approximate magnitude of Hoover Dam and Power plant at a fraction of the cost and impact on the environment when compared to any other means of providing new generation capacity.

Low-head Hydropower

A low-head dam is one with a water drop of less than 65 feet and a generating capacity less than 15,000 kW. Large, high-head dams can produce more power at lower costs than low-head dams, but construction of large dams may be limited by lack of suitable sites, by environmental considerations, or by economic conditions. In contrast, there are many existing small dams and drops in elevation along canals where small generating plants could be installed. New low-head dams could be built to increase output as well. The key to the usefulness of such units is their ability to generate power near where it is needed, reducing the power inevitably lost during transmission.

Peaking with Hydropower

Demands for power vary greatly during the day and night. These demands vary considerably from season to season, as well. For example, the highest peaks are usually found during summer daylight hours when air conditioners are running.

Nuclear and fossil fuel plants are not efficient for producing power for the short periods of increased demand during peak periods. Their operational requirements and their long startup times make them more efficient for meeting base load needs.

Since hydroelectric generators can be started or stopped almost instantly, hydropower is more responsive than most other energy sources for meeting peak demands. Water can be stored overnight in a reservoir until needed during the day, and then released through turbines to generate power to help supply the peak load demand. This mixing of power sources offers a utility company the flexibility to operate steam plants most efficiently as base plants while meeting peak needs with the help of hydropower. This technique can help ensure reliable supplies and may help eliminate brownouts and blackouts caused by partial or total power failures.

Today, many of Reclamation=s 58 power plants are used to meet peak electrical energy demands, rather than operating around the clock to meet the total daily demand. Increasing use of other energy-producing power plants in the future will not make hydroelectric power plants obsolete or unnecessary. On the contrary, hydropower can be even more important. While nuclear or fossil-fuel power plants can provide base loads, hydroelectric power plants can deal more economically with varying peak load demands. This is a job they are well suited for.

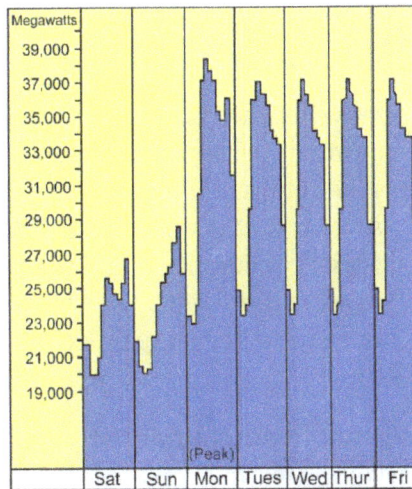

Typical Weekly Load Curve of a Large Electric Utility

Pumped Storage

Like peaking, pumped storage is a method of keeping water in reserve for peak period power demands. Pumped storage is water pumped to a storage pool above the power plant at a time when customer demand for energy is low, such as during the middle of the night. The water is then allowed to flow back through the turbine-generators at times when demand is high and a heavy load is place on the system.

The reservoir acts much like a battery, storing power in the form of water when demands are low and producing maximum power during daily and seasonal peak periods. An advantage of pumped storage is that hydroelectric generating units are able to start up quickly and make rapid adjustments in output. They operate efficiently when used for one hour or several hours.

Because pumped storage reservoirs are relatively small, construction costs are generally low compared with conventional hydropower facilities.

Conduit Hydroelectricity

Hydroelectric power generation is classified as "conduit hydropower" if the water flowing through the generating equipment has a primary use other than for power generation. For example, any power generation equipment installed on a pipeline or similar man-made water conveyance system that is operated for the distribution of water for agricultural, municipal or industrial consumption and not primarily for the generation of electricity can be classified as a conduit hydropower installation. This classification is significant because it does not require licensing or exemption by the Federal Energy Regulatory Commission (FERC).

Numerous applications for conduit hydropower exist within water systems, including pressure-reducing sites, reservoirs, treatment plants, industrial processes and irrigation. The main benefit of conduit hydropower experienced by water system operators is reduced operational costs. The power generated by a conduit hydroelectric installation can be sold to the power utility at a wholesale rate ranging anywhere from $0.04/kWh up to $0.20/kWh, depending on the prevailing cost of power at the location. If power is used at a site where electricity is consumed by a pump or other equipment, it can be "net metered." This can reduce both demand charges and the monthly electrical bill, which lowers overall operating costs. Conduit hydropower is a continuous source of energy that typically follows the same diurnal curve as power demand thereby helping to level out the power grid supply and demand. Power utilities generally prefer this type of generation over more intermittent forms of generation such as solar or wind.

Making a Site Suitable

To generate power and offset operating costs, a potential site needs to have both a pressure differential and a volume flow rate. The pressure differential available for power generation is typically the difference in height from the intake to the discharge or in the case of a control valve application, it is simply the pressure differential across the valve. If the differential pressures and flow rates are known, a conduit hydroelectric firm can provide an estimate of how much energy and revenue can be generated at the site.

A suitable site for conduit hydroelectric installations are those with high-pressure differentials and high continuous flows because they will generate the most power. However, variable flow, fixed flow and even intermittent flow applications are often cost effective as well. The potential site should have nearby power lines for interconnection. Conduit projects do not need to be new installations; they can utilize existing infrastructure such as existing vaults, piping and valving to reduce installation costs.

For example, a pressure-reducing valve that has a pressure drop of 50psi and an average flow rate of 2,000gpm could generate up to 35kW of power. If this valve operated year round it would generate 275,000kWh per year. If this power was then sold at a wholesale rate of $.10/kWh, approximately $27,500 per year of revenue would be generated. If this generation was installed at a pump station or treatment plant, for example, a net metering arrangement could be set up in which power generated would directly offset power purchased at the retail rate.

Typical System Configuration

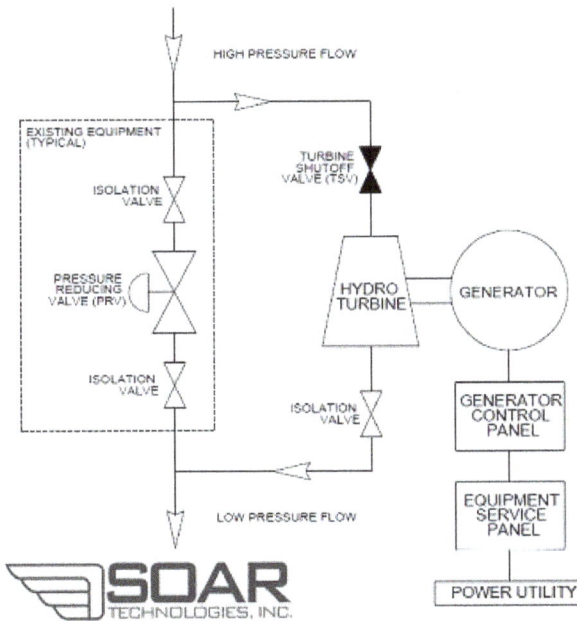

Water delivery is always the main priority of a water distribution system; therefore, conduit hydroelectric projects are configured to be a secondary priority to water delivery. When generating equipment is coupled with a custom control system, the hydro turbine functions in the same way as a control valve. Modes of operation can including pressure reducing, pressure sustaining, level control and flow control among others. The hydro turbine is designed similarly to a control valve in that it relies on inputs from the water system or SCADA to function properly. An experienced hydropower firm will take care to thoroughly understand water system operations before designing and installing a conduit hydroelectric system.

A typical hydroelectric system is plumbed parallel to a new or existing control valve. This parallel valve becomes the turbine bypass valve. When the turbine is down for maintenance or emergency, water delivery can continue uninterrupted. An automatic fail-safe valve is usually installed on the high-pressure side of the hydro turbine to operate as the turbine shutoff valve. The turbine shutoff valve allows the generating equipment to be safely shut down in the event of a power outage or emergency situation. The turbine control system automatically and seamlessly transfers the flow of water from the hydro turbine to the bypass control valve and back upon startup and shutdown.

Head and Flow

Head Hydropower all comes down to head and flow. The amount of power, and therefore energy that you can generate is proportional to the head and the flow.

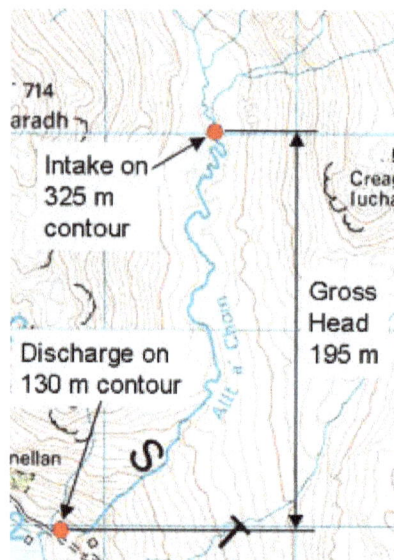

Diagram of measuring head at high head hydropower site

Head is the change in water levels between the hydro intake and the hydro discharge point. It is a vertical height measured in metres. The two diagrams below show how the head would be measured on a typical 'low head' and a typical 'high head' site. The more head you have the higher the water pressure across the hydro turbine and the more power it will generate. Higher heads are not only better because they generate more power, but also because the higher water pressure means you can force a higher flow rate through a smaller turbine, and because turbine cost is closely related to physical size, higher-head turbines often cost less than their low-head cousins even though they might generate the same power.

Higher head also means a faster rotating turbine and generator, which means lower torque.

The cost of drive train is closely related to how much torque it has to transmit, so higher heads = less torque = less cost.

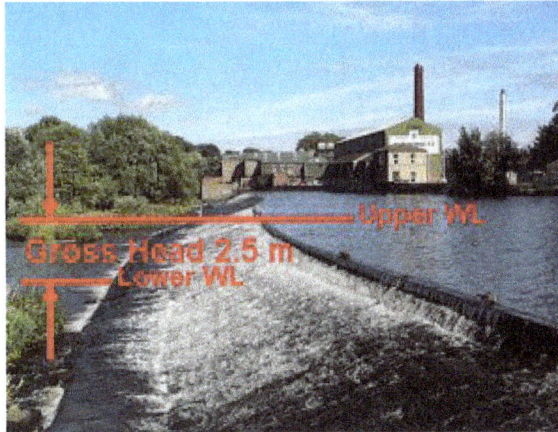

Diagram of measuring head at a low head hydropower site

Of course you only have what you have, so if your site only has 2 ½ metres of head you won't be able to increase this significantly. However, even small increases in head can make a difference.

Sometimes it is possible to clear silt or re-grade a tailrace or discharge channel to lower the downstream water levels slightly which increases the overall head at the site. Or it may be possible to raise the water level on the upstream side by raising weir crests or sluices, though this must be done carefully to avoid increasing flood risk, and sometimes requires the construction of new spillways or installation of fail-safe tilting weirs to ensure that flood risk isn't increased during extreme flood events.

Generally speaking the cost of even small increases in head at low-head sites is repaid hundreds of times over from increased energy production for the next few decades, so is always worth the effort.

Flow

The flow rate and how it varies over a year is the next and equally important parameter.

The simplest way to characterise the flow in a watercourse is to work out what the *long-term annual mean flow* is. This is important because it is the overall average flow in a watercourse that is important; it doesn't matter if it is a raging torrent after heavy rains because in the big scheme of things we only have really heavy rains for a few days or weeks a year, and for the other 50 weeks when it is lightly raining, drizzling or bright summer sunshine you would still want you hydro system to be working and generating energy.

The fundamental piece of information that characterises the flow in a watercourse is the *flow duration curve*. Although simple once you understand them, they are quite complicated to newcomers. figure below shows the long-term flow duration curve of a small river in Somerset.

River Yeo Daily Mean Flow
Data set 1/5/1964 to 30/4/2002

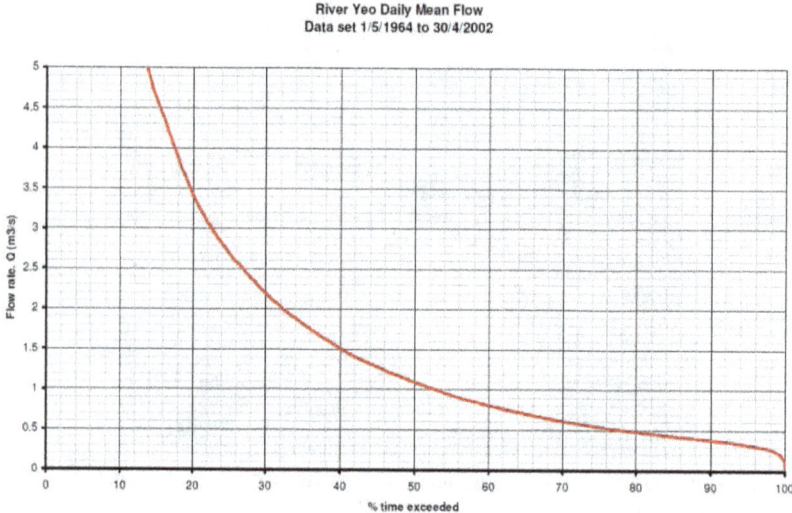

Figure: Long-term flow duration curve for the River Yeo in Somerset

The y-axis is the flow rate in m³/s (metres-cubed per second), or sometime in litres per second for smaller watercourses. When the flow duration curve is constructed all of the flow rate data is sorted into descending order, then the highest flow rates are plotted on the left of the curve, then progressively lower flow rates to the right until the very lowest flow is plotted at the extreme left-hand end. The x-axis is the 'percentage exceedence. This is normally the difficult part to understand. For a given percentage exceedence it shows the flow rate equalled *or exceeded* for that percentage of time. For example, if you look at the 50% percentage exceedence on figure above and read off the flow rate at that point you will see that it is 1.1 m³/s. This doesn't mean that the flow rate in the watercourse is 1.1 m³/s for 50% of the year, it means that the flow rate is 1.1 m³/s *or more* for 50% of the year. This is more important because it is clear from the shape of the curve that apart from the instant that the line crosses the 50% mark it is always more than 1.1 m³/s.

The x-axis is always plotted as a percentage exceedence from 0 to 100%. This is so that any data set spanning any interval can be plotted. The data set may span 10 days or more likely several decades of data from a local river gauging station. Generally speaking flow duration curves present long-term annual data, so the flow rates read off them are the annual flow characteristics. The percentage exceedences are often called 'Q values', so Q_{95} is the flow rate exceeded for 95% of a year and Q_{10} the flow exceeded for 10% of the year. The data set can also be analysed using a spreadsheet to work out the average (arithmetic mean) of all of the flows, and this is called Q_{mean}. In the UK the Q_{mean} is normally somewhere between Q_{25} and Q_{30}, and in the case of figure above is $Q_{26.5}$.

Once you understand how a flow duration curve is constructed it can tell you a great deal about the flow characteristics of the water course. Firstly you can work out the Q_{mean}, which is the average flow, and often if you are negotiating with the environmental regulator you will be discussing the Q_{95} flow, as this is often used as the representative 'low flow' that must always flow in any depleted stretches of river while the main volume of water passes through the hydro turbine.

By comparing different Q values you can see how 'flashy' a watercourse is, or whether it has a high base flow. The flow duration curve in figure has a Q_{mean} of 2.54 m³/s and a Q_{95} of 0.32 m³/s, so the Q_{95} is 8% of the Q_{mean}. This is typical for a 'flashy' river that rises and falls quickly in response to rain because the rain runs straight of the land and into the river without getting stored in bogs or porous rocks before flowing into the river sometime later.

If you make the same comparison using figure, you will see that the Q_{mean} is 11.30 m³/s and the Q_{95} 5.83 m³/s. In this case Q_{95} is 52% of the Q_{mean}. This would be a 'high base-flow' river, typical of the rivers that flow in Hampshire with chalk catchments where the chalk stores the rain in its porous structure and then releases it to the rivers slowly and over a long period of time, rather like a sponge. It is also interesting to note that Q_{mean} is at Q_{41}; much lower on the flow duration curve than a flashy river like in figure above.

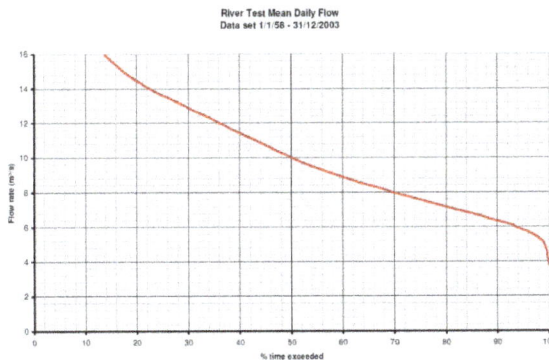

Long-term flow duration curve for the River Test in Hampshire

In both flow duration curves you can see that the red line is heading steeply upwards at the left-hand end of the graph. This is the 'extreme flow' region, which is not much use for energy generation because such extreme flows occur for a relatively small proportion of the year, but they are very important when designing a hydropower systems to ensure that the hydro structures and turbine house don't get flooded, or even worse, washed away during the next major storm.

Minimum Head and Flow Required

For a commercially viable site it would normally need to be at least 25 kW maximum power output. For a low-head micro hydropower system you would need at least 2 metres of gross head and an average flow rate of 2.07 m³/s. To put this in context this would be a small river that was approximately 7 metres wide and around 1 metre deep in the middle.

For a site with 25 metres head a much lower average flow rate of 166 litters/second would be needed. This would be a large stream of 2 – 3 metres width and around 400 mm deep in the middle.

It is technically possible to develop smaller hydropower sites with lower power outputs, but the economics start to get challenging. This is particularly true for low-head sites; when the head drops to 1.5 metres it isn't normally possible to get any kind of return on investment, though the site could still be technically developed using Archimedean screws or modern waterwheels.

The table below shows the average flow rates needed for a range of heads from 2 metres to 100 metres for system with 100, 50, 25, 10 and 5 kW maximum power outputs. 25 kW would normally be considered the minimum for a commercial project, though a 10 kW system can still produce an acceptable return if the civil engineering works are simple (hence don't cost much). 5 kW systems are not normally viable, but the figures are shown for interest and may be useful for sites that can generate value from non-tangible benefits such as attracting visitors or positive publicity.

	Maximum Power Output (kW)				
	5	10	25	50	100
Head (m)	Flow required (m³/sec)				
2	0.340	0.680	1.699	3.398	6.796
5	0.136	0.272	0.680	1.359	2.718
10	0.068	0.136	0.340	0.680	1.359
50	0.014	0.027	0.070	0.136	0.272
100	0.006	0.014	0.034	0.068	0.136

Minimum flow rates required for a range of (gross) heads.

Hydropower Potential

For any stretch of a watercourse, characterized by a difference in level of H meters, conveying a discharge of Q (m³ /sec), the theoretical (potential) power,

$$N_p = \gamma QH = 1000QH \, (\text{kgm}/\sec)$$

$$N_p = \frac{1000}{75} QH = 13.3QH = (HP)$$

$$N_p = 13.3 \times 0.736 QH = 9.8QH \, (kW)$$

If the rate of flow changes along a stretch, the mean value of the discharges pertaining to the two terminal sections of the stretch is to be substituted in the equation,

$$Q = (Q_1 + Q_2)/2,$$

The theoretical power resources of any river or river system are given by the total of the values computed for the individual stretches,

$$N_p = \sum \frac{1000QH}{75} \times 0.736 = 9.8 \sum QH \, (kW)$$

Potential water power resources can be characterized by different values according to the discharge taken as basis of computation. The conventional discharges are,

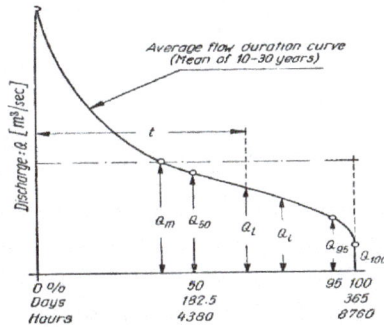

Figure: Discharges used for characterising potential water power resources

1. Minimum potential power, or theoretical capacity of 100%, is the term for the value computed from the minimum flow observed. N_{p100}.

2. Small potential power. The theoretical capacity of 95% can be derived from the discharge of 95% duration as indicated by the average flow duration curve. N_{p95}.

3. Median or average potential power. The theoretical capacity of 50% can be computed from the discharge of 50% duration as represented by the average flow duration curve. N_{p50}.

4. Mean potential power. The value of theoretical mean capacity can be ascertained by taking into account the average of mean flow. The average of mean flow is understood as the arithmetic mean of annual mean discharges for a period of 10 to 30 years. The annual mean discharge is the value that equalizes the area of the annual flow duration curve.

Economic Significance of Potential Water Resources of a Site

This is influenced by a great number of factors than hydraulic, such as geographical, geological and topographical conditions, energy demand, etc. Ignoring these and comparing relative values of power potential as reflected by hydraulic conditions only, the following four aspects are to be taken into consideration:

a) The absolute quantity of theoretical water power resources.

b) The relative share of discharge in the power. Among the hydraulic possibilities representing equal magnitudes, the more advantageous are those where the power in question originates from a smaller flow and a higher head. It is advantageous of highland developments over power stations situated in hilly regions or lowland areas.

c) The relative annual fluctuation of available potential power.

This can be characterized by the ratio of the values Np_{50} to the values $Np95$ (or Np_{100}).

$$\alpha = \frac{N_{p50}}{N_{p95}} \text{ or, } \quad \alpha_1 \frac{N_{p50}}{N_{p100}}$$

A smaller ratio reflects a more favorable hydraulic possibility.

d) The over year or multi-annual variation of potential power.

This can be characterized either by a simple diagram showing the annual potential power against time, or by a summation curve of annual values.

Power resources can be characterized even by annual values of potential energy in a river, by the quantities of work,

$$E_{100}, E_{95}, E_{50} \text{ and } E_m$$

All expressed in kilowatt-hours. These values can be computed as areas of the lower parts of the potential power-duration curves. If the head is assumed to be constant, independent of discharge, the computation can be based on the discharge-duration curve. Using the curve,

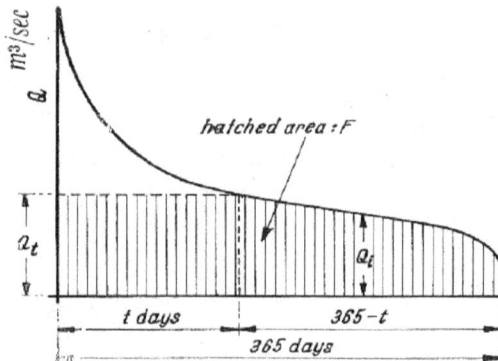

$$N_p = 9.8 H \alpha Q_1 = \alpha Q_1 (kW)$$

$$E = 24 \alpha Q_1 t + 24 \sum_t^{365} \alpha Q_1$$

$$E = 24 \alpha \left(Q_1 t + \sum_t^{365} Q_1 \right) = 24 \alpha F (kWh)$$

Where,

t = the duration considered in days,

Q_t = Selected discharge,

Q_i = Daily mean of actual discharge at any time,

F = Area pertaining to Q_t (shaded area).

The upper limit of potential energy inherent in the river section is obtained by,

$$E_{mx} = (24 \times 365) N_m = 8760 N_m (kWh)$$

Where N_m is the annual mean power.

The overall coefficient is about 0.75 or 0.80. The equation recommended is,

$$N_{mnet} = (7.4 - 8.0) \sum_t^{365} Q_m H$$

Q_m = the arithmetic mean discharge. Net river energy potential,

$$E_{max} = 8760 \sum N_{m\,net}$$

For characterizing the gross potential power of a river basin, the following data should be used:

a) Total annual discharge volume V (m³),

b) Medium height of the watershed area H lying upstream above sea level (m),

c) Area A of this basin (km²).

Since 1 m³ water weighs 1 ton, the product VH yields the annual gross power pertaining to the selected site in meter-tons,

$$E(kWh) = \frac{VH}{367}$$

$$e(kWh / km^2) = \frac{VH}{367 A}$$

If the e values are determined along the river basin at several stream sites, then the lines connecting the points of equal e, isopotential lines can be drawn.

Flow-Duration Curve

If the flows for any unit time are arranged in descending order of time (without regard to chronological sequence), the percentage of time for which any magnitude is equaled or exceeded may be computed. The resulting array is called a flow-duration curve.

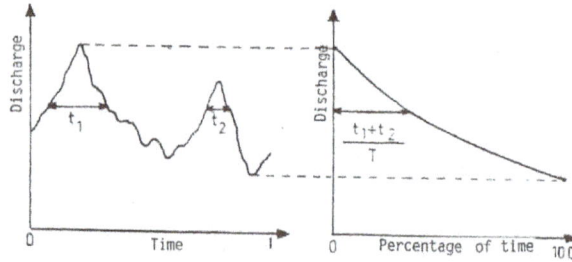

Figure: Determination of flow-duration curve

Such curves are useful in determining the relative variability of flow between two points in a river basin or between two basins. For example, if a stream is highly regulated, the curve will approach a horizontal line. The dependable flow is that corresponding to 100 percent of time. The relative variability of two flow records may be compared by converting the discharge scale in terms of a ratio to the mean. Any sub area under the curve represents the volume of annual runoff.

Flow-duration curves have been used to approximate the amount of storage needed to increase the dependable flow. For example, the horizontal line AB in below figure may represent a new dependable flow, and the required storage needed to obtain this flow is indicated by area ABC.

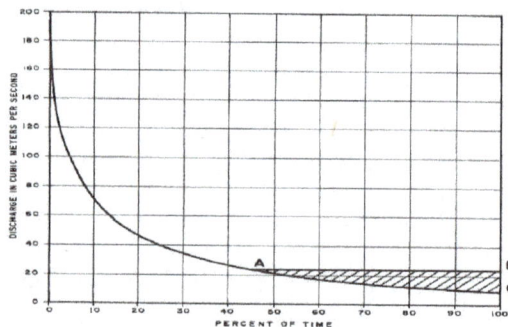

Figure: Duration curve of monthly discharges.

Power production values may be approximated from the duration curve by converting the discharge scale to kilowatts by multiplying by a selected head, efficiency and conversion factors. If the time scale is converted to hours in a year, a unit of are represents kilowatt-hours.

The flow-duration curve is particularly useful in combination with a sediment rating curve (river discharge versus the transported sediment load usually expressed in tons per day), to compute total sediment load to be expected in an average year.

Flow Mass Curve

Total flow volume from a certain initial time t = 0 up to time t can be computed as,

$$H = \int_{0}^{t} Qdt$$

In practice, the total volume is computed as,

$$H = \sum Q_i t_i$$

Q_i = the average discharge in time interval (month, year) Δt_i.

Flow mass curve is a plot of the cumulative runoff from the hydrograph against time. The time scale is the same as for the hydrograph and may be in days, months or years. The volume ordinate may be in m³ -days, m³ -months, m³ -years, etc. The slope of the mass curve is the derivative of the volume with respect to time or the rate of discharge.

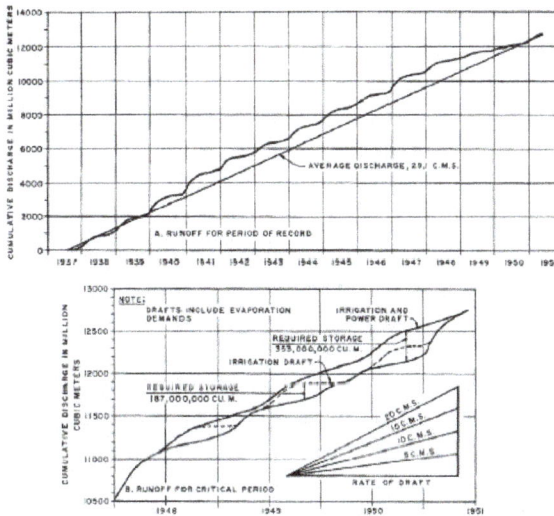

The mass curve usually has a wavy configuration in which the steeper segments represent high flow periods and flatter segments represent low flows. Uniform rates of withdrawal (draft) may be represented as tangent lines drawn from high points to intersect the curve at the next wave. The vertical distance between the draft line and the basic curve represents the cumulative difference between regulated outflow and natural inflow, or the required storage. If the draft line does not intersect the mass curve at the end of a year, it means that the reservoir does not refill with that rate of draft and regulation at the proposed draft rate will extend over two years or more. A typical mass curve is shown in the above figure.

In estimating storage requirements from the mass curve, it is not necessary to assume a constant rate of regulated flow. For example, if the draft rate to meet a demand for irrigation, water supply, or power varies from month to month, the draft line may be a curved or irregular line and the maximum draft may not occur at the low point in the mass curve.

An allowance for evaporation should be applied to the mass curve analysis. If the water area does not change significantly during the annual cycle of use, an average correction for each calendar month can be subtracted from the inflow or added to the draft rates.

The ordinates of the flow mass curve increase continuously in time. The sum of the differences between the inflow and the yield (average flow) are drawn;

$$H_o = \sum (Q_i - Q_{ave}) \Delta t_1$$

Reservoir capacity is then vertical distance between the highest and lowest points of the curve.

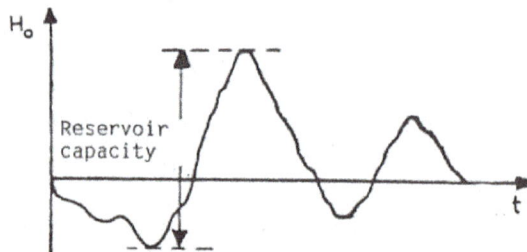

Figure: Flow mass curve derived using the differences
of the discharges from the yield.

Storage-Draft Curve

The results of a mass-curve analysis can be plotted as a storage-draft curve. His curve gives the storage needed to sustain various draft rates. Examples of storage-draft curves are shown in the below figures. Both irrigation requirements and combined irrigation and power requirements are illustrated. These curves were computed from the mass curve.

If storage unlimited, the storage-draft curve will approach the available mean flow as asymptote. It is rarely possible to develop mean annual flow of a river basin. For most projects, some spillage will occur in years of runoff. To impound all flood flows will require an extensively large reservoir. Such a reservoir may not fill in many years, and probably could not be justified economically. The selected rate of regulated flow to be developed will depend on:

1. The demand of water users,

2. The available runoff,

3. The physical limits of the storage capacity,

4. The overall economies of the project.

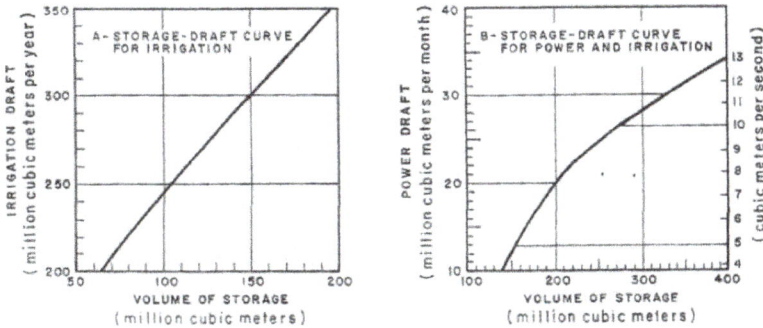

Figure: Storage-draft curves for multipurpose uses

Selection of Design Flow

The hydrologic analyses, combined with economic analyses of costs and benefits for different heights of dam and reservoir capacity will lead to the selection of the reservoir capacity and the corresponding dependable flow that can be justified. The selected design flow may not necessarily be available 100 percent of the time. The propose water use may permit deficiencies at intervals, for example, a 15 percent shortage once in 10 years. Irrigation water supplies may permit greater deficiencies than those for urban and industrial use. Hydroelectric power plants, connected to large systems, may tolerate substantial water supply deficiencies.

Final Storage Selection

a) Evaporation Losses

Detailed evaluation of evaporation losses should be postponed until final operation and routing studies, when the actual variation in water area can be considered as well as the seasonal variation in evaporation.

Basic data on water surface evaporation may be obtained from records of pan evaporation. Such records overestimate lake evaporation and must be reduced by a pan coefficient which varies from 0.60 to 0.80 depending on the climate. The collection of evaporation records at a project site should be initiated in the planning stage. Evaporation corrections should be made on a monthly basis using actual past precipitation records at the project site if possible.

b) Power

Selection of an average flow alone will not permit determination of the benefits from a water resources development project without more detailed studies. Such studies require routing through the reservoir the entire record of flow (corrected for evaporation losses), on a month by month basis, using assumed patterns of use, outlet capacities and, in the case of power, turbine and generator capacities and efficiencies. The reservoir would normally be considered to be full at the start of the operation study, or at least full to normal pool.

For power benefits, the energy output will vary in accordance with the inflow, outflow, and change in storage and corresponding head, tail water elevation, turbine capacity and plant efficiency. If the plant is a part of a system, the output may be subject to varying demands of the system load curve and whether the plant is to be used as a base load plant or a peaking plant. The routing study will indicate the necessary modifications to the head, storage, and even height of the dam to obtain maximum benefit.

c) Irrigation

Operation studies for irrigation use should be made using seasonal crop demands and selected outlet capacities. Short-term demands may indicate that the storage needed war greater than that required for uniform regulated flow. The proposed annual water use may be greater than that available 100 percent of the time, with the understanding that deficiencies can be tolerated in some years.

d) Water Supply

Operation studies for projects providing urban water supplies will be similar to those for irrigation projects in that there may be variations in the seasonal demand, especially where more than one source is available, or where there can be transfers to other regulation reservoirs. However, the degree of dependability of flow must be higher for urban water supply than for irrigation projects.

e) Flood Control

The storage allocated for flood control in single purpose or multipurpose projects is usually based on a definite design flood the control of which is needed for downstream protection. The required storage capacity is based on routing of the design flood inflow coincident with releases not to exceed downstream channel capacities.

Total Storage Requirement

The total usable storage needed for multipurpose for multipurpose projects require more complex routing studies and numerous trials to obtain the most economic allocations.

In addition to the variable requirement for storage for downstream uses, the total storage may be increased for the following reasons:

- Minimum head on power installations.

- Allowance for the storage of sediments without loss of usable storage.

- Minimum area for recreation use, including seasonal requirements.

Example: Monthly flow volumes feeding a reservoir are given in the table. Determine the storage capacity required to supply the mean annual flow volume yield.

Solution: Cumulative volumes are calculated and given in the table,

Month	Volume ($10^6 m^3$)	Cumulative Volume ($10^6 m^3$)
1	296	296
2	386	682
3	504	1186
4	714	1900
5	810	2710
6	1154	3864
7	746	4610
8	1158	5768
9	348	6116
10	150	6266
11	223	6489
12	182	6671

Total volume of flow feeding the reservoir is 6671×106 m³. Annual mean discharge can be calculated as,

$$Q = \frac{6671 \times 10^6}{365 \times 86400} \cong 212 \ m^3/s$$

The reservoir storage capacity required to obtain 212 m³ /s yield throughout the year is found by drawing tangents parallel to the average draft line from peak points. The vertical distance is 1800×106 m³ is the required capacity of the reservoir.

The reservoir capacity to supply the annual mean discharge can be found out by using the sum of differences method as in table,

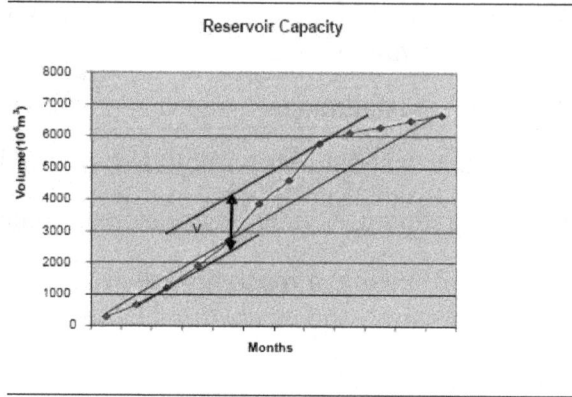

Reservoir Capacity

Month	Volume ($10^6 m^3$)	Flow (m^3/s)	Ho (m^3/s)	ΣHo (m^3/s)
1	296	111	-101	-101
2	386	149	-63	-165
3	504	188	-24	-188
Month	Volume (106m3)	Flow (m3/s)	Ho (m3/s)	ΣHo (m3/s)
1	296	111	-101	-101
2	386	149	-63	-165
3	504	188	-24	-188

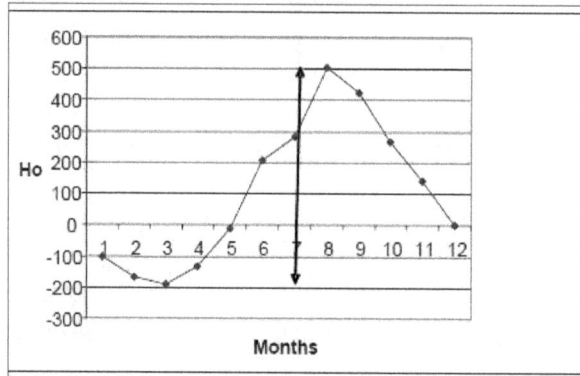

$$H_o = \sum (Q_i - Q_{ave}) \Delta t_1$$

Reservoir Capacity $= 504 - (-188) = 693 \, m^3 / s$

Volume $= 692 \times 31 \times 86400 = 1853 \times 10^6 \, m^3$

Reservoirs

A reservoir is a manmade lake or structure used to store water. A dam reservoir has an uncontrolled inflow but a largely controlled outflow. The water available for storage is totally a function of the natural stream stream-flow that empties into it.

Reservoir Capacity

Reservoir capacity is the volume of water that can be stored in the particular reservoir. It is the normal maximum pool level behind a dam. This can be calculated by using a topographic map of the region. First, the area inside different elevation contours is measured, and then a curve of area versus elevation can be constructed.

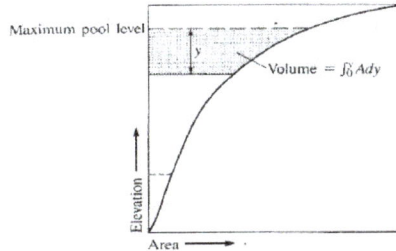

Figure: Area versus elevation for a reservoir

At any given elevation, the increment of storage in the reservoir at that elevation will be Ady, where dy is a differential depth. Then the total storage below the maximum level to any will be given by,

$$V = \int_{0}^{y} A d y$$

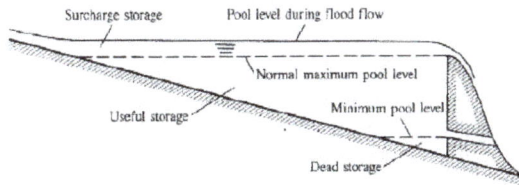

Figure: Storage relations for a reservoir with an uncontrolled spillway

Sedimentation in Reservoirs

All streams carry sediments that originate from erosion processes in the basins that feed the streams. After a dam is constructed across the steam and a reservoir is produced, the velocity in the reservoir will be negligible so that virtually all the sediment coming into the reservoir will settle down and be trapped. Therefore, the reservoir should be designed with enough volume to hold the sediment and still operate as a water storage reservoir over the project's design life. For large projects, the design life is often considered 100 years.

Sediment carried in a stream is classified as either *bed load* or *suspended load*. The bed load consists of the coarsest fractions of the sediment (sands and gravels), and it rolls, slides, and bounces along the bottom of the stream. The finer sediments are suspended by the turbulence of the stream. When the sediment enters the lower velocity zone of the reservoir, the coarser sediments will be deposited first, and it is in this region that a delta will be formed. The finer sediments will be deposited beyond the delta at the bottom of the reservoir.

The total sediment outflow from a watershed or drainage basin measured in a specified period is the *sediment yield*. The yield is expressed in terms of tons per square kilometer per year. The engineer designing a reservoir must estimate the average sediment yield for the basin supplying the reservoir to determine at what rate the reservoir will fill with sediment.

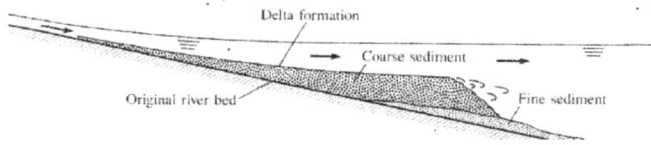

Figure: Deposition of sediment in a reservoir

For a given reservoir volume, V, the ratio of the reservoir volume decrease due to the deposited sediment can be estimated by this empirical equation,

$$R = 23 \times 10^{-6} G^{0.95} \left(\frac{A}{V} \right)^{0.8}$$

Where,

G = Sediment yield of the basin (kN/km^2/year)   A = Drainage basin area (m^2)

V = Reservoir volume (m^3)

Multiplying R ratio with the design life of the reservoir, T, will yield the percentage of the dead volume in the reservoir. The dead volume can be estimated over the period of design life by,

$$V_{dead} = R \times T \times V_{reservoir}$$

EXAMPLE: The total volume of a reservoir is V = 230×10^6 m^3 with a drainage basin of A = 1200 km^2. The design life of the project is T = 100 year and the density (specific mass) of the deposit is ρ = 2.65 ton/m^3. Calculate the dead volume of the reservoir.

Solution: The sediment of the river for a river,

$$G = 1421 A^{-0.229}$$
$$G = 1421 \times 1200^{-0.229}$$
$$G = 280 \left(m^3/km^2/year \right) (\text{volume})$$

The ratio of the reservoir volume decrease every year,

$$R = 0.000023 \times G^{0.95} \left(\frac{A}{V} \right)^{0.8}$$

$$R = 0.000023 \times 7279^{0.95} \times \left(\frac{1200 \times 10^6}{230 \times 10^6} \right)^{0.8}$$

$$R = 0.4 \,(o/o)$$

Reservoir volume decrease due to the sediment deposit every year,

$$V_{dead} = 0.004 \times 230 \times 10^6 = 0.92 \times 10^6 \ m^3$$

For the 100 year of design period,

$$V_{dead100} = 100 \times 0.92 \times 10^6 = 92 \times 10^6 \ m^3$$

Useful storage is,

$$V_{useful} = (230 - 92) \times 10^6 = 138 \times 10^6 \ m^3$$

Wind-Generated Waves, Setup and Freeboard

Whenever wind blow over an open stretch of water, waves develop, and the mean level of the water surface may change. The latter phenomenon, called *setup* or *wind tide*, is significant only in relatively shallow reservoirs. When a dam is designed, the crest of the dam must be made higher than the maximum pool level in the reservoir to prevent overtopping of the dam as the wind-generated waves strike the face of it. The additional height given to the crest of the dam to take care of wave action, setup, and possibly settlement of the dam (if it is earth fill) is called *freeboard*.

Setup

Consider the basin of water shown in the figure. The solid line depicting the water surface is the case when no wind is blowing; the water surface is horizontal. When the wind is blowing, a shear stress acts on the water surface, and because of this, the surface will tilt, as shown by the broken line in the below figure.

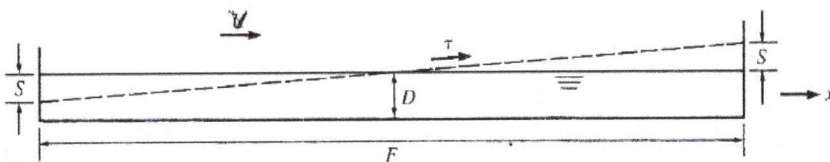

Figure: Definition sketch for setup

The amount of setup S is,

$$S = \frac{V^2 F}{KD}$$

Where,

D = Average reservoir depth (m),

V = the wind speed measured at a height of 10 m from the surface (km/h) F = the wind fetch (km)

K = A constant $\approx$ 62000

S = Setup (m)

EXAMPLE: A reservoir is oval shaped with a length of 20 km and a width of 10 km. If the wind blows in a direction lengthwise to the reservoir with a velocity of 130 km/h, what will be the setup of the average water depth of the reservoir is 10 m.

Solution: The setup will be,

$$S = \frac{V^2 F}{KD}$$
$$= \frac{130^2 \times 20}{62000 \times 10} = 0.55m$$

Height of Wind Waves and the Run-Up

Allowances for wave height and the run-up of wind-generated waves are the most significant components of freeboard. The *run-up* of the waves on the upstream dam face, i.e. the maximum vertical height attained by a wave running up a dam face, is equal to H (wave height) for a typical vertical face in deep water, but can attain values over 2H for a smooth slope 1 in 2.

A *wave height*, H (m), (crest to trough) can be estimated by,

$$H = 0.34\sqrt{F} + 0.76 - 0.26\sqrt[4]{F}$$

Where,

F = the fetch length (km),

H = Wave height (m)

For large values of fetch (F>20 km), the last two terms may be neglected. With the provision for the wind speed, the equation takes the form of,

$$H = 0.32\sqrt{VF} + 0.76 - 0.26\sqrt[4]{F}$$

Where,

V = Wind velocity (km/hour)

The freeboard will be equal to set-up plus run up allowance for settlement of the embankment plus and amount of safety (usually 0.50m).

EXAMPLE: Calculate the wind set-up and wave height for a reservoir with 8 km fetch length. The average reservoir depth is 15 m. The wind velocity is V = 100 km/h. If the upstream of the dam is vertical, what will be the minimum freeboard to be given?

Solution:

The wind set-up is,

$$S = \frac{V^2 F}{KD}$$

$$= \frac{100^2 \times 8}{62000 \times 15} = 0.9m$$

The wave height,

$$H = 0.32\sqrt{VF} + 0.76 - 0.26\sqrt[4]{F}$$

$$H = 0.032\sqrt{100 \times 8} + 0.76 - 0.26\sqrt[4]{8}$$

$$H = 1.19m$$

Since the upstream side of the dam is vertical, the run-up height will be taken as the height of the wave. The freeboard,

$$H_{freeboard} = 0.09 + 1.19 + 1.19 + 0.5$$

$$H = 2.97m \cong 3.00m$$

0.5 m is the safety height.

Tidal Energy Generation

Tidal energy or tidal power is a form of renewable energy obtained due to alternating sea levels. The kinetic energy from the natural rise and fall of tides is harnessed and converted into electricity. Tides are caused by the combined gravitational forces of the moon, sun, and earth. However, tides are influenced most by the moon. The moon's gravitational force is so strong that it tugs the ocean into bulge. The high and low tides create tidal currents, which are essential in generation of this kind of energy mostly prevalent in coastal areas. Tidal energy is considered a renewable energy resource because the oceans and seas will remain until the end of time and tides are highly predictable.

Tidal energy generation plants are most commonly installed along coastlines although offshore plants are increasingly becoming popular. Coastlines are preferred because they receive 2 high tides and 2 low tides every single day. To generate electricity, the disparity in water levels must be at least 5 meters.

Tidal power has great potential for future as tides can be much more accurately predicted than wind or sun and due to massive size of oceans. Though it available in plenty but harnessing energy from it is not that easy. It suffers from huge investment and limits availability of sites where it can be captured.

Conversion of Tidal Energy into Electricity

Tidal energy is converted into electricity using three main tidal technologies:

1. Tidal Turbines

Tidal turbines utilize the same technology to wind turbines. The only difference is that the blades of tidal turbines are way stronger and shorter. So, the best way to compare tidal turbines is underwater windmills. Ideally, the water currents turn the turbine. The turbine is connected to a generator through a shaft. So, when the turbine turns, the shaft also turns. The turning shaft activates a generator, which generates electricity. The initial cost of setting up this tidal stream system is quite on the higher side, not to mention the difficulty in maintenance. However, it remains a cheaper alternative and doesn't cause environmental degradation compare to other tidal technologies.

2. Tidal Barrages

Tidal Barrages are the most efficient tidal energy technologies. They resemble dams used in hydropower plants. The difference is they are a lot bigger since they are constructed across a Bay or an Estuary. For the barrage to be able to produce power, the tidal range, which is the difference between low and high tide, has to be more than 5 meters. As the tide enters the system, ocean or sea water flows via the dam into the basin. When the tides subside, the system's gates close, trapping the water in the estuary or basin. When the tides start to move out, the gates in the dam that consist of turbines, open up, and water begins to flow out hitting the turbines, which eventually turn to produce energy. Construction of tidal barrages involves high upfront capital costs, plus they have devastating effects on the local environment.

3. Tidal lagoons

This technology has a lot in common with tidal barrages. It's just that it doesn't involve a lot of initial capital outlay and it's friendly to the environment. Tidal lagoon is a power station separated from the rest of the ocean or sea. Its functionality is similar to tidal barrage since when the tide goes up, the lagoon completely fills up. When the tide subsides, the water is allowed to drain out through an opening consisting of turbines. The outward flow of water turns the turbine, which generates energy.

Advantages of Tidal Energy

By taking the time to critically analyze the advantages and disadvantages of tidal energy, you can get a clear picture of whether it's a sustainable energy resource or not. Below is an outline of the upsides and downsides to tidal energy.

1. It's environmental friendly

The fact that tidal energy technologies are installed on the coastlines and offshore makes them good for the environment since land will not be interfered with. Also, tidal energy is a clean source of energy, meaning it doesn't release any greenhouses gasses to the atmosphere.

2. It's a renewable energy

Tides harnessed to produce tidal energy result from the combined gravitational pull of the sun, moon, and earth in conjunction with the rotation of the planet around its axis. This is a natural process that occurs every single day. This means that tides will continue to occur and production of tidal energy will continue until the end of time.

3. It's highly predictable

Development of tides is a well-understood cycle. This makes it a lot easier to develop tidal energy systems with the right dimensions. Why? Because the level of power the system will be exposed to is already determined. Which explains why the capacity of the installed equipment and the entire physical size has completely no energy generation limitations, even though stream generators and tidal turbines used resemble those of wind turbines.

4. It's Cost-competitive

Tidal energy technologies once constructed have the potential to generate electricity for many years, which means they are long lasting. Although the upfront costs of setting up a tidal power plant are relatively high, the return on investment will be realized in the long run. A typical example is the La Rance tidal barrage, which is still producing electricity since 1966.

5. It Minimizes over-dependence on fossil fuels

Fossil-based sources of energy such as oil, coal and natural gas emit greenhouse gasses that lead to climate change and global warming. Tidal energy offers a green and renewable alternative to cut back on greenhouse gas emissions.

6. It Offers a sense of protection

Barrages and dams that are utilized to tap tidal energy for the generation of electricity could insulate coastal areas and ship ports from high impact and dangerous tides in the course of bad weather and storms.

7. It's quite effective than wind even at low speeds

Oceans currents have the capacity to produce more energy than air currents because ocean water is 832 times denser than air. This means ocean currents applies greater force on the turbines to generate more energy.

Disadvantages of Tidal Energy

1. High upfront capital costs

Tidal energy technologies are considerably new. Meaning the costs of infrastructure are relatively high at the moment. Experts also project that tidal energy will only start to be commercially beneficial in 2020 with enhancements of innovative technologies.

2. completely environmentally friendly

Tidal energy generation systems are thought to harbor some environmental impacts, but they have not been quantified. In addition, these tidal plants produce electricity using tidal barrages that depend on manipulation of sea levels. This means that they have the same environmental impacts as hydroelectric dams.

3. Problems of efficiency

Generation of tidal electricity wholly depends on tidal surges, which happen twice a day. This means when tides are not happening, there is no production of energy, which is why extra costs must be incurred to set up energy storage systems.

4. Tidal energy needs Long gestation period

Tidal power plants need a lot of time to be able to produce electricity efficiently. This aspect combined with cost of installation can be unsustainable. A typical example of a tidal power plant that was closed due to time and cost overruns is the UK's Severn Barrage.

5. Impact on marine life

The greatest fear among tidal energy systems developers is the impact the plants and turbines will have on the surrounding marine ecosystem. The rotation of turbines and vibrations of the tidal plant could significantly interrupt marine ecosystem and inhibit natural movement of marine life.

Tidal Current

Tidal currents, as their name suggests, are generated by tides. Tides are essentially long, slow waves created by the gravitational pull of the moon, and to a lesser degree, the sun, on the earth's surface. Since the moon is so much closer to the earth than the sun, its pull has more influence on the tides.

The moon's gravitational pull forces the ocean to bulge outwards on opposite sides of the earth, which causes a rise in the water level in places that are aligned with the moon and a decrease in water levels halfway between those two places. This rise in water level is accompanied by a horizontal movement of water called the tidal current.

Tidal currents also switch directions every time the tide transitions between high and low. Although tides and tidal currents don't have much impact in the open oceans, they can create a rapid current of up to 15.5 miles (25 kilometers) per hour when they flow in and out of narrower areas like bays, estuaries and harbors. Fast tidal currents toss sediment around and affect plant and animal life. Currents may, for example, transfer a fish's eggs from an estuary out into the open sea or carry nutrients that the fish needs from the sea into the estuary.

The strongest tidal currents occur at or around the peak of high and low tides. When the tide is rising and the flow of the current is directed towards the shore, the tidal current is called the flood current, and when the tide is receding and the current is directed back out to sea, it is called the ebb current. Because the relative positions of the moon, sun and earth change at a known rate, tidal currents are predictable.

Tidal Stream

Tidal Stream is the name given to the horizontal flow of water through the oceans caused by the continuous ebb and flood of the tide, which as we know is the vertical up-down movement of the oceans water. Unlike water currents which are a continuous, unidirectional and form a steady horizontal movement of water flowing down a river or stream etc. a *tidal stream* or tidal current, changes its speed, direction and horizontal movement regularly according to the forces of the tide controlling it.

Tidal stream generation is a non-barrage tidal scheme, unlike tidal fence energy which uses a physical barrier to extract the energy. Tidal stream systems extract the kinetic energy (energy in motion) from moving water generated by the tides without altering the environment thereby making it a hydrokinetic energy system.

At or near the coast, the ebb and flood of the tides causes the oceans waters to pile up resulting in a high tide along the beach, with some of this water being forced into tidal inlets, basins and estuaries while the majority is forced sideways along the shore. This movement of the tidal range amplified by geographical features along the coastline, focuses these tidal currents into a single predictable and concentrated form of renewable energy which we can exploit using a tidal stream generator. A tidal stream is usually stronger nearer to the coast where the sea water is naturally shallower causing the water to speed, than it is farther out in deeper depths.

Tidal Stream Generator

Tidal Stream Generation is very similar in many ways to the principles of wind power generation. Horizontal turbine generators called "tidal turbines" or "marine current turbines" are placed on the ocean floor, the stream currents flow across the turbine blades powering a generator much like how wind turns the blades of wind power turbines. In fact, in some tidal stream generation areas the sea bed looks just like underwater wind farm with arrays of tidal stream generators covering large areas.

The generated tidal electricity is then transmitted to the shore via long underwater electrical cables called *submarine cables*. These offshore tidal turbines can be either partially or fully submerged beneath the surface of the water, with partially submerged turbines being easier and less costly for maintenance.

While tidal stream installations reduce some of the environmental effects of large man-made tidal barrages, major ocean currents like the Gulf Stream, travel at speeds significantly slower than the wind. However, as water is 784 times more dense than air (which is why we can see water and not air), a single tidal generator sitting on the sea bed can provide a significant amount of ocean current energy at low tidal stream velocities which is far superior to wind, using similar or identical turbine technology.

Since energy output varies with the density of the medium, (Kg/cm^3) and the cube of the velocity, (m^3/s), we can see that a 10 mph (about 8.6 knots in nautical terms) ocean tidal current would have an energy output equal or greater than a 90 mph wind speed for the same size of turbine system. Therefore, even small increases in velocity can lead to substantial changes in the amount of available power and therefore, smaller faster rotating tidal turbine generators can be used in a ocean based tidal stream system.

As the kinetic energy content of a tidal stream flows per unit time, which is the same as the hydro power (P), the available energy can be calculated in terms of velocity (V), swept cross-sectional area (A) perpendicular to the stream flow direction, and the density of the water (ρ), which for sea water is approximately 1025 kg/m^3. Providing the velocity is uniform across the cross-sectional area, at any instant in the tidal cycle the amount of energy available will be: $P = \frac{1}{2} \rho A.V^3$.

This cubic relationship between velocity and power is the same as that for the power curves relating to wind turbines, but there are practical limits to the amount of power that can be extracted from tidal streams. Some of these limits relate to the design of the tidal stream turbines and the characteristics of the underwater resource.

Tidal stream generators make use of the kinetic energy of moving water to turn a turbine - similar to the way a wind turbine uses wind to create electricity. However, the power available for tidal power generation in a given area can be greater than a wind turbine due to the higher density of water. These types of tidal generators tend to be the cheapest (though still quite expensive) and most environmentally friendly of any type of tidal power generation. There are several different specific types of tidal stream generators.

These types of generators have a very low visual impact and are mainly, if not totally submerged. Furthermore, they are less intrusive to marine life because they produce less noise pollution.

Tides are more predictable, than wind which varies a great deal. The density of water is also much higher than air, which means that tidal turbines can be much smaller than wind turbines of comparable output. This is further enhanced by the use of the Venturi effect, which is a way of getting water to move faster through these turbines.

As with all types of tidal power technology the initial investment costs are extremely high and it takes quite a while to make back that investment. Unfortunately, sea water is quite corrosive which leads to high maintenance costs.

Since the tides flow-in and recede once each per day tidal stream generators provide intermittent power generation.

Varying types of generators

Different designs of tidal power generation may prove to offer different advantages, extensive studies are still needed on this relatively new technology. Studies are needed for maintenance costs vs. income and and environmental impacts over an extended period of time. Power output can vary considerably between these classes of turbines. However, since the tidal power industry is so new predictions for power output vary considerably. Currently, research data on stream generators is scarce due to the emerging nature of the industry.

Axial (Horizontal) Turbines

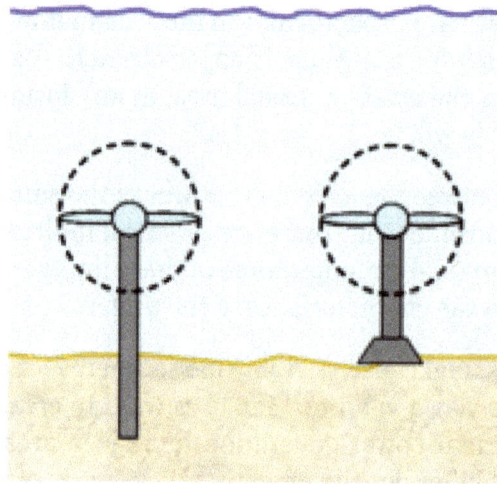

Figure: Axial turbines underwater.

Axial turbines have a rotor that is parallel to the incoming water stream. They use rotors similar to those seen on wind turbines but have modifications due to the differing fluid properties of water from air.

These devices use the lift of water to generate power. This requires specially shaped airfoil surfaces designed to create a pressure difference. This leads to a net force in the direction perpendicular to the water flow and thus turns the device. Rotors of this type must be carefully oriented (the orientation is referred to as the rotor pitch), to maintain their ability to harness the power of the tide as it changes.

Cross-flow Turbines

Cross flow turbines have a rotor that is perpendicular to the water flow but parallel to the water surface, see figure above Unlike axial turbines, these devices typically use the drag of water to generate power. In drag-based turbines, the force of the water pushes against a surface, like wind on an open sail. This works because the drag of the open face of the turbine blade is greater than the drag on its closed face. Drag based devices are inherently less efficient than their lift based counterparts because they operate with respect to the relative flow speed of the water. This means that as a turbine rotates faster the relative flow speed of the water will decrease and thus transfer less of its energy.

Figure: A cross flow water turbine.

Vertical axis Turbines

With vertical axis turbines the rotational axis of the rotor is vertical to the water surface and also perpendicular to the incoming water flow. Both lift and drag type blades can be used in these turbines.

Oscillating Generators

Oscillating devices do not have a rotational component, instead these generators make use of sections which are pushed sideways by the flowing current to create a hydraulic pump. This pump transfers its energy to a motor, which then turns a generator and creates electricity.

Tidal Stream Generator Designs

Unlike off-shore wind power which can suffer from storm or heavy sea damage, tidal or marine current turbines operate just below the sea surface or are permanently fixed to the sea bed. Most submerged tidal turbines essentially operate in the same way as a wind turbine and are fastened to the ocean floor, with water pushing the turbine instead of the wind. These turbines have an axis of rotation horizontal to the ground and operates like a traditional windmill consisting of a rotor, a gearbox, and an electrical generator. These three parts are mounted onto a steel support structure with the three main types of support being a gravity structure, a sunken piled structure or a tripod structure as shown,

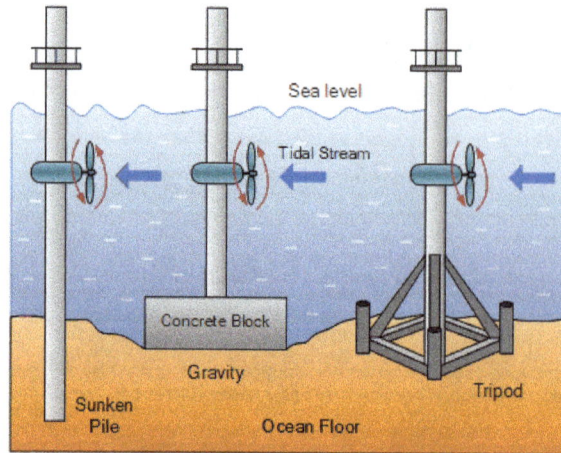

Tidal Stream Generator Supports

For a sunken pile support, a single steel pile is driven deep into the sea bed with the tidal stream generator assembly attached to it. This tubular support is less stiff than other types and can flex under the downstream drag forces of the tidal waters when used in shallow waters. A gravity support generally uses a large heavy concrete block or blocks which sit on the sea bed. Due to the heavy weight of the concrete block, the structure is stiffer and therefore more resistant to flexing. A tripod or truss support uses a tubular frame with a much larger footprint positioned on the ocean floor to support the generator assembly. This type of system is used in oil and gas exploration so is a known technology.

Other tidal stream generator designs fixed to the ocean floor include: Reciprocating Tidal Stream Devices that uses a large hydrofoil similar to a whales flipper, which moves up and down parallel to the direction of the tidal stream instead of rotating blades, and Venturi Effect Tidal Stream Devices, were the tidal turbines are located inside a cylindrical duct, much like a fan housing. The tidal flow is funnelled through this duct, which concentrates the flow producing a pressure difference causing a secondary water flow through the reaction turbine thereby improving efficiency.

Also, there are several practical advantages in placing the tidal turbine inside a fan type duct, such as less dangers from the rotating blades to both aquatic marine life and divers as a safety grill or cover could be placed on the upstream opening which would also have the secondary advantage of preventing floating debris from being drawn or sucked into the turbine causing damage. The duct itself can provide shading and/or shelter for the reaction turbine from direct sunlight, preventing seaweed, algae growth or crustaceans forming on the blades and mechanism as they do on the underside of boats.

We know that tidal streams are formed by the fast flowing horizontal currents of water caused by the ebb and flow of the tides with the profile of the sea bed causing the water currents to speed up, or slow down near the shoreline. Then tidal stream turbines can generate power on both the ebb and flow of the tide. One of the disadvantages of Tidal

Stream Generation is that, as the turbines are submerged under the surface of the water they can create hazards to large sea mammals, navigation and shipping.

Sea Barge

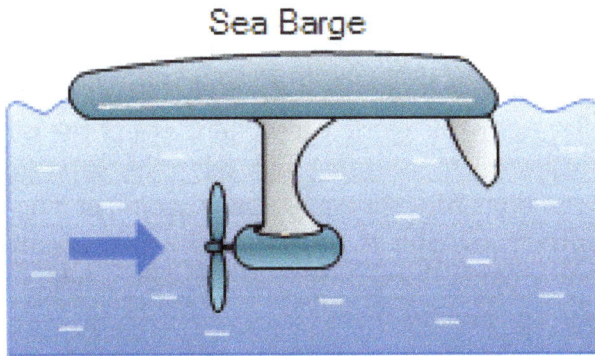

Given the technical difficulties resulting from underwater corrosion, increased maintenance issues, weed growth on the blades, which could reduce their efficiency and stability concerns, other forms of alternative tidal stream generator designs are now being used. These include the tidal turbine being connected to a floating barge or ship on the waters surface, essentially operating as an upside down horizontal turbine instead of fastening the turbines directly to the ocean floor.

There are numerous advantages to this type of tidal stream generator design, including easy maintenance and accessibility of the turbines, by simply removing or replacing them out of the water, and no costly steel supports or alterations to the ocean floor. Also, as the tidal turbines are located under a barge, pontoon or fixed directly to the hull of a ship, they can have their electrical connections and equipment mounted safely above and out of the water. Plus the supporting flotation device can be easily moved to stronger tidal stream areas if required, but they are limited by distance due to their umbilical electrical cable connected to the shoreline.

Tidal Barrage

The Tidal Barrage or Tidal Power Plant as it is also known is a form of "marine renewable energy" generation system that uses long walls, dams, sluice gates or tidal locks to capture and store the potential energy of the ocean. A *Tidal Barrage* is a type of tidal power generation scheme that involves the construction of a fairly low walled dam, known as a "tidal barrage" and hence its name, spanning across the entrance of a tidal inlet, basin or estuary creating a single enclosed tidal reservoir, similar in many respects to a hydroelectric impoundment reservoir.

The bottom of this barrage dam is located on the sea floor with the top of the tidal barrage being just above the highest level that the water can get too at the highest annual tide. The barrage has a number of underwater tunnels cut into its width allowing the sea water to flow through them in a controlled way by using "sluice gates" on their entrance and exit points. Fixed within these tunnels are huge tidal turbine generators that spin as the sea water rushes past them either to fill or empty the tidal reservoir thereby generating electricity.

The water which flows into and out of these underwater tunnels carries enormous amounts of kinetic energy and the job of the tidal barrage is to extract as much of this energy as possible which it uses to produce electricity. Tidal barrage generation using the tides is very similar to hydroelectric generation, except the water flows in two directions rather than just one. On incoming high tides, the water flows in one direction and fills up the tidal reservoir with sea water. On outgoing ebbing tides, the sea water flows in the opposite direct emptying it. As a tide is the vertical movement of water, the tidal barrage generator exploits this natural rise and fall of tidal waters caused by the gravitational pull of the sun and the moon.

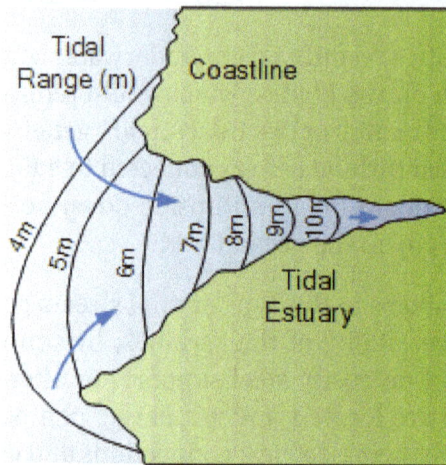

Effects of Tidal Flow in an Estuary

The gravitational effects of the sun or the moon on the worlds oceans causes huge amounts of sea water to be directed towards the nearest coastline. The result of this movement of water is a rise in the sea level. In the open ocean, this rise is very small as there is a large surface area with deeper depths for it to flow into.

However, as the oceans water moves nearer towards the coastline, the sea level rises steeply especially around inlets and estuaries because of the upward sloping gradient of the sea bed. The effect of this sloping gradient is to funnelling the water into the estuaries, lagoons, river inlets and other such tidal "bottlenecks" along the coastline.

The result of funnelling all of this water is that the height of the sea level once inside these inlets can increase vertically many metres every day as it is being pushed forward by the incoming sea water behind it as shown in the image. This increase in the sea level can create a tidal range of over ten metres in height in some estuaries and locations which can be exploited to generate electricity.

The tidal range is the vertical difference between the high tide sea level and the low tide sea level. The tidal energy extracted from these tides is potential energy as the tide moves in a vertical up-down direction between a low and a high tide and back to a low creating a height or head differential. A tidal barrage generation scheme exploits this head differential to generate electricity by creating a difference in the water levels either side of a dam and then passing this water difference through the turbines. The three main tidal energy barrage schemes that use this water differential to their advantage are:

- Flood Generation: in which the tidal power is generated as the water enters the tidal reservoir on the incoming tide.

- Ebb Generation: in which the tidal power is generated as the water leaves the tidal reservoir on the ebb tide.

- Two-way Generation: in which the tidal power is generated as the water flows in both directions during a flood and ebb tide.

Tidal Barrage Flood Generation

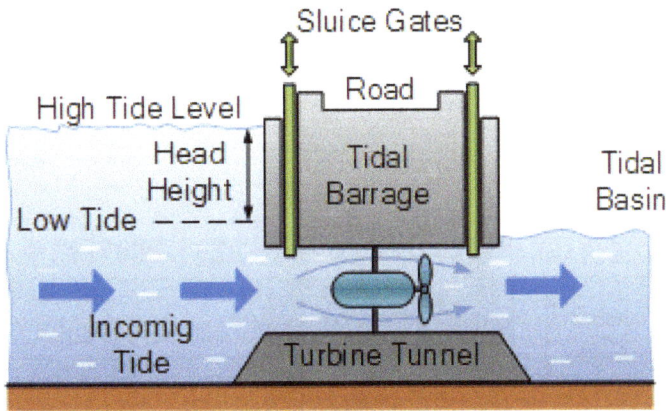

A Tidal Barrage Flood Generation uses the energy of an incoming rising tide as it moves towards the land. The tidal basin is emptied through sluice gates or lock gates located along a section of the barrage and at low tide the basin is affectively empty. As the tide turns and starts to comes in, the sluice gates are closed and the barrage holds back the rising sea level, creating a difference in height between the levels of water on either side of the barrage dam.

The sluice gates to the entrances to the dams tunnels can either be closed as the sea water rises to allow for a sufficient head of water to develop between the sea level and the basin level before being opened generating more kinetic energy as the water rushes through, turning the turbines as it passes. Or may remain fully open, filling up the basin more slowly and maintaining the same water level inside the basin as out in the sea.

The tidal reservoir is therefore filled up through the turbine tunnels which spin the turbines generating tidal electricity on the flood tide and is then emptied through the opened sluice or lock gates on the ebb tide. Then a flood tidal barrage scheme is a one-way tidal generation scheme on the incoming tide with tidal generation restricted to about 6 hours per tidal cycle as the basin fills up.

The movement of the water through the tunnels as the tidal basin fills up can be a slow process, so low speed turbines are used to generate the electrical power. This slow filling cycle allows for fish or other sea life to enter the enclosed basin without danger from the otherwise fast rotating turbine blades. Once the tidal basin is full of water at high tide, all the sluice gates are opened allowing all the trapped water behind the dam to return back to the ocean or sea as it ebbs away.

Flood generator tidal power generates electricity on a incoming or flood tide, but this form of tidal energy generation is generally much less efficient than generating electricity as the tidal basin empties, called "Ebb Generation". This is because the amount of kinetic energy contained in the lower half of the basin in which flood generation operates is much less the kinetic energy present in the upper half of the basin in which ebb generation operates due to the effects of gravity and the secondary filling of the basin from inland rivers and streams connected to it via the land.

Tidal Barrage Ebb Generation

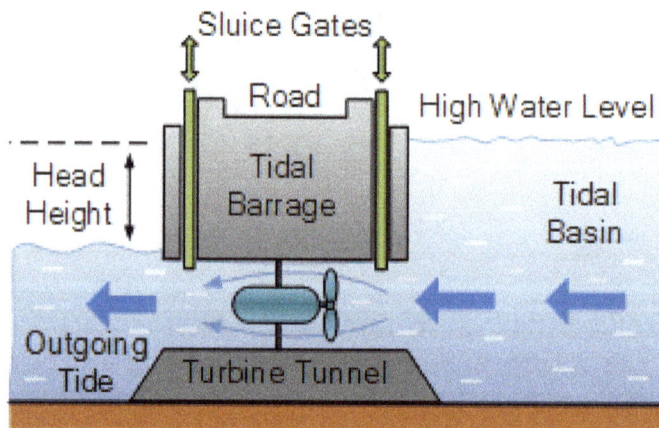

A Tidal Barrage Ebb Generation uses the energy of an outgoing or falling tide, referred to as the "ebb tide", as it returns back to the sea making it the opposite of the previous flood tidal barrage scheme. At low tide, all the sluice and lock gates along the barrage are fully opened allowing the tidal basin to fill up slowly at a rate determined by the incoming flood tide.

When the ocean or sea level feeding the basin reaches its highest point at high tide, all the sluices and lock gates are then closed entrapping the water inside the tidal basin (reservoir). This reservoir of water may continue to fill-up due to inland rivers and streams connected to it from the land.

As the level of the ocean outside the reservoir drops on the outgoing tide towards its low tide mark, a difference between the higher level of the entrapped water inside the tidal reservoir and the actual sea level outside now exists. This difference in vertical height between the high level mark and the low mark is known as the "head height".

At some time after the beginning of the ebb tide, the difference in the head height across the tidal barrage between the water inside the tidal reservoir and the falling tide level outside becomes sufficiently large enough to start the electrical generation process and the sluice gates connected to the turbine tunnels are opened allowing the water to flow.

When the closed sluice gates are opened, the trapped potential energy of the water inside flows back out to the sea under the enormous force of both gravity and the weight of the water in the reservoir basin behind it. This rapid exit of the water through the tunnels on the outgoing tide causes the turbines to spin at a fast speed generating electrical power.

The turbines continue to generate this renewable tidal electricity until the head height between the external sea level and the internal basin is too low to drive the turbines at which point the turbines are disconnected and the sluice gates closed again to prevent the tidal basin from over draining and effecting local wildlife. At some point the incoming flood tide level will again be at a sufficient level to open all the lock gates filling-up the basin and repeating the whole generation cycle over again as shown.

Power Generation during Ebb Tide

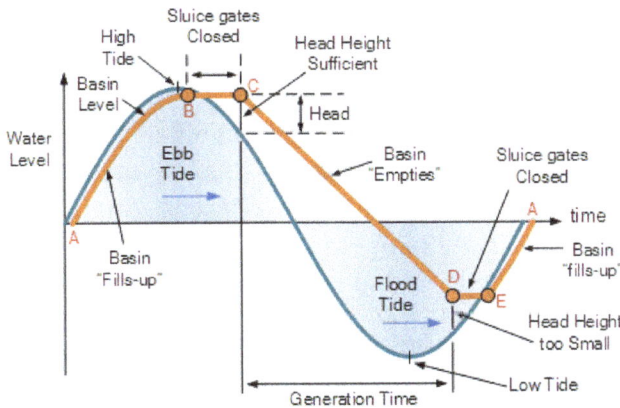

From the graph above we can see that the tidal basin fills up between points E and B via A on the incoming flood tide. Once high tide is reached, the sluice gates are then closed between points B – C to stop the tidal basin from emptying. Therefore no power is generated between points E – C while the basin is filling up.

When there is sufficient head height between either side of the barrage, the sluice gates are opened at point C releasing the trapped water back to the sea and power generation starts. The tidal basin continues to empty driving the tidal generators until the level of water in the basin reaches point D.

At point D the head height of the sea water across the barrage is no longer adequate to drive the turbines so the sluice gates are closed and generation stops until the level of the sea water reaches point E again and the whole process starts over. Then Ebb Tidal Barrage Generation also known as outgoing generation, takes its name because the electrical generation only occurs as the tide ebbs or flows out.

At point E the sea level becomes sufficient to re-fill the basin and the sluice gates are opened. Therefore, the tidal reservoir is filled up through the sluice or lock gates on the flood tide and is emptied through the turbine tunnels which spin the turbines generating tidal electricity on the ebb tide. Then an ebb tidal barrage scheme is a one-way tidal generation scheme which operates between points C – D on the graph above.

Two-way Tidal Barrage Generation Scheme

We have seen above that both Flood Tidal Barrage and Ebb Tidal Barrage installations are "one-way" tidal generation schemes, but in order to increase the power generation time and therefore improve efficiency, we can use special double effect turbines that generate power in both directions. A Two-way Tidal Barrage Scheme uses the energy over parts of both the rising tide and the falling tide to generate electricity.

Two-way electrical generation requires a more accurate control of the sluice gates keeping them closed until the differential head height is sufficient in either direction before being opened. As the tide ebbs and flows, sea water flows in or out of the tidal reservoir through the same gate system. This flow of tidal water back and forth causes the turbine generators located within the tunnel to rotate in both directions producing electricity.

However, this two-way generation is in general less efficient than one-way flood or ebb generation as the required head height is much smaller which reduces the period over which normal one-way generation might have otherwise occurred. Also, bi-directional tidal turbine generators designed to operate in both directions are generally more expensive and less efficient than dedicated unidirectional tidal generators.

One way of improving the operating time and efficiency of a two-way tidal barrage scheme is to use individual one-way unidirectional tidal turbines inverted along the barrage. By controlling their individual sluice gates, one set of turbines can be made to operate on the flood tide between points E - B and the other set operate on the ebb tide between points C - D. While this method increases the total number of tidal turbines located along the tidal barrage, it has the advantage that the generation period is greatly extended.

Pumping

Turbines are able to be powered in reverse by excess energy in the grid to increase the water level in the basin at high tide (for ebb generation). Much of this energy is returned during generation, because power output is strongly related to the head. If water is raised 2 ft (61 cm) by pumping on a high tide of 10 ft (3 m), this will have been raised by 12 ft (3.7 m) at low tide.

Environmental Impact

The placement of a barrage into an estuary has a considerable effect on the water inside the basin and on the ecosystem. Many governments have been reluctant in recent times to grant approval for tidal barrages. Through research conducted on tidal plants, it has been found that tidal barrages constructed at the mouths of estuaries pose similar environmental threats as large dams. The construction of large tidal plants alters the flow of saltwater in and out of estuaries, which changes the hydrology and salinity and possibly negatively affects the marine mammals that use the estuaries as their habitat. The La Rance plant, off the Brittany coast of northern France, was the first and largest tidal barrage plant in the world. It is also the only site where a full-scale evaluation of the ecological impact of a tidal power system, operating for 20 years, has been made.

French researchers found that the isolation of the estuary during the construction phases of the tidal barrage was detrimental to flora and fauna, however; after ten years, there has been a "variable degree of biological adjustment to the new environmental conditions"

Some species lost their habitat due to La Rance's construction, but other species colonized the abandoned space, which caused a shift in diversity. Also as a result of the construction, sandbanks disappeared, the beach of St. Servan was badly damaged and high-speed currents have developed near sluices, which are water channels controlled by gates.

Turbidity

Turbidity (the amount of matter in suspension in the water) decreases as a result of smaller volume of water being exchanged between the basin and the sea. This lets light from the Sun penetrate the water further, improving conditions for the phytoplankton. The changes propagate up the food chain, causing a general change in the ecosystem.

Tidal Fences and Turbines

Tidal fences and turbines, if constructed properly, pose less environmental threats than tidal barrages. Tidal fences and turbines, like tidal stream generators, rely entirely on the kinetic motion of the tidal currents and do not use dams or barrages to block channels or estuarine mouths. Unlike barrages, tidal fences do not interrupt fish migration or alter hydrology, thus these options offer energy generating capacity without dire environmental impacts. Tidal fences and turbines can have varying environmental impacts depending on whether or not fences and turbines are constructed with regard to the environment. The main environmental impact of turbines is their impact on fish.

If the turbines are moving slowly enough, such as low velocities of 25-50 rpm, fish kill is minimalized and silt and other nutrients are able to flow through the structures For example, a 20 kW tidal turbine prototype built in the St. Lawrence Seaway in 1983 reported no fish kills Tidal fences block off channels, which makes it difficult for fish and wildlife to migrate through those channels. In order to reduce fish kill, fences could be engineered so that the spaces between the caisson wall and the rotor foil are large enough to allow fish to pass through Larger marine mammals such as seals or dolphins can be protected from the turbines by fences or a sonar sensor auto-braking system that automatically shuts the turbines down when marine mammals are detected.

Salinity

As a result of less water exchange with the sea, the average salinity inside the basin decreases, also affecting the ecosystem. "Tidal Lagoons" do not suffer from this problem.

Sediment movements

Estuaries often have high volume of sediments moving through them, from the rivers to the sea. The introduction of a barrage into an estuary may result in sediment accumulation within the barrage, affecting the ecosystem and also the operation of the barrage.

Fish

Fish may move through sluices safely, but when these are closed, fish will seek out turbines and attempt to swim through them. Also, some fish will be unable to escape the water speed near a turbine and will be sucked through. Even with the most fish-friendly turbine design, fish mortality per pass is approximately 15% (from pressure drop, contact with blades, cavitation, etc.). Alternative passage technologies (fish ladders, fish lifts, fish escalators etc.) have so far failed to solve this problem for tidal barrages, either offering extremely expensive solutions, or ones which are used by a small fraction of fish only. Research in sonic guidance of fish is ongoing. The Open-Centre turbine reduces this problem allowing fish to pass through the open centre of the turbine.

Recently a run of the river type turbine has been developed in France. This is a very large slow rotating Kaplan-type turbine mounted on an angle. Testing for fish mortality has indicated fish mortality figures to be less than 5%. This concept also seems very suitable for adaption to marine current/tidal turbines.

Energy calculations

The energy available from a barrage is dependent on the volume of water. The potential energy contained in a volume of water is:

$$E = \tfrac{1}{2} A \rho g h^2$$

Where:

- h is the vertical tidal range,

- A is the horizontal area of the barrage basin,

- ρ is the density of water = 1025 kg per cubic meter (seawater varies between 1021 and 1030 kg per cubic meter),

- g is the acceleration due to the Earth's gravity = 9.81 meters per second squared.

The factor half is due to the fact, that as the basin flows empty through the turbines, the hydraulic head over the dam reduces. The maximum head is only available at the moment of low water, assuming the high water level is still present in the basin.

Example Calculation of Tidal Power Generation

Assumptions:

- The tidal range of tide at a particular place is 32 feet = 10 m (approx)

- The surface of the tidal energy harnessing plant is 9 km^2 (3 km × 3 km)= 3000 m × 3000 m = 9×10^6 m^2

- Density of sea water = 1025.18 kg/m^3

Mass of the sea water = volume of sea water × density of sea water.

$\quad$ = (area × tidal range) of water × mass density

$\quad$ = (9×10^6 m^2 × 10 m) × 1025.18 kg/m^3

$\quad$ = 92×10^9 kg (approx)

Potential energy content of the water in the basin at high tide = ½ × area × density × gravitational acceleration × tidal range squared.

$\quad$ = ½ × 9×10^6 m^2 × 1025 kg/m^3 × 9.81 m/s^2 × (10 m)^2

$\quad$ =4.5×10^{12} J (approx)

Now we have 2 high tides and 2 low tides every day. At low tide the potential energy is zero. Therefore, the total energy potential per day = Energy for a single high tide × 2.

$\quad$ = 4.5×10^{12} J × 2

$\quad$ = 9×10^{12} J

Therefore, the mean power generation potential = Energy generation potential / time in 1 day.

$= 9 \times 10^{12}$ J / 86400 s

$= 104$ MW

Assuming the power conversion efficiency to be 30%: The daily-average power generated = 104 MW * 30%.

$= 31$ MW (approx)

Because the available power varies with the square of the tidal range, a barrage is best placed in a location with very high-amplitude tides. Suitable locations are found in Russia, USA, Canada, Australia, Korea, the UK. Amplitudes of up to 17 m (56 ft) occur for example in the Bay of Fundy, where tidal resonance amplifies the tidal range.

Wave Power

Wave Energy also known as Ocean Wave energy, is a type of ocean based renewable energy source that uses the power of the waves to generate electricity.

Unlike tidal energy which uses the ebb and flow of the tides, *wave energy* uses the vertical movement of the surface water that produce tidal waves. Wave power converts the periodic up-and-down movement of the oceans waves into electricity by placing equipment on the surface of the oceans that captures the energy produced by the wave movement and converts this mechanical energy into electrical power.

Wave energy is actually a concentrated form of solar power generated by the action of the wind blowing across the surface of the oceans water which can then be used as a renewable source of energy. As the sun rays strike the Earth's atmosphere, they warm it up. Differences in the temperature of the air masses around the globe causes the air to move from the hotter regions to the cooler regions, resulting in winds.

As the wind passes over the surface of the oceans, a portion of the winds kinetic energy is transferred to the water below, generating waves. In fact, the ocean could be viewed as a vast storage collector of energy transferred by the sun to the oceans, with the waves carrying the transferred kinetic energy across the surface of the oceans. Then we can say that waves are actually a form of energy and it is this energy and not water that moves along the ocean's surface.

These waves can travel (or "propagate") long distances across the open oceans with very little loss in energy, but as they approach the shoreline and the depth of the water becomes shallower, their speed slows down but they increase in size. Finally, the wave crashes onto the shoreline, releasing an enormous amount of kinetic energy which can be used for electricity production. A breaking waves energy potential varies from place to place depending upon its geographic location and time of year, but the two main factors which affect the size of the wave energy are the winds strength and the uninterrupted distance over the sea that the wind can blow.

Then we can say that "Wave Energy" is an indirect form of wind energy that causes movement of the water on the surface of the oceans and by capturing this energy the motion of the waves is converted to mechanical energy and used to drive an electricity generator. In many respects, the technology used for capturing this wave energy is similar to tidal energy or hydroelectric power.

The kinetic energy of the wave turns a turbine attached to a generator, which produces electricity. However, the open oceans can be a stormy and violent environment, resulting in the wave energy machines being destroyed by the very energy they were designed to capture.

In its simplest terms, an ocean wave is the up-and-down vertical movement of the sea water which varies sinusoidally with time. This sinusoidal wave has high points called crests and low points called troughs. The difference in height of a wave between the crest and the trough is called the peak-to-peak amplitude, then the waves amplitude or height is the centre of these two points and corresponds to the actual sea level when there is no movement of the water, in other words, a calm sea.

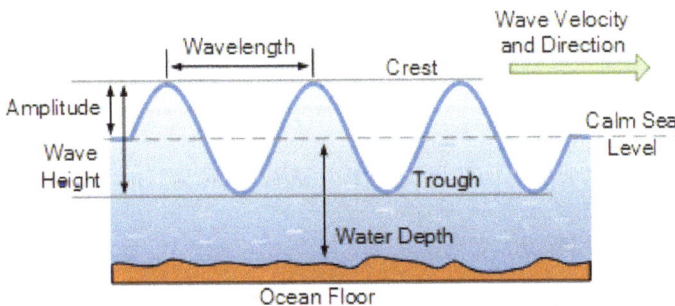

The amplitude of an ocean wave depends on the weather conditions at that time, as the amplitude of a smooth wave, or swell, will be small in calm weather but much larger in stormy weather with strong gales as the sea water moves up and down.

As well as the amplitude of the wave, another important characteristic is the distance between each successive crest, or trough, known as the wave period, (T). This wave period is the time in seconds between each crest of the wave. Then for a gentle swell this time period may be very long, but for a stormy sea this time period may be very short as each wave crashes onto the one in front.

The reciprocal of this time (1/T) gives us the fundamental frequency of the ocean wave relative to some static point. Smaller periodic waves generated or superimposed onto this fundamental wave such as reflected waves are called harmonic waves. Then the frequency and amplitude characteristics of a wind-generated wave depend on the distance the wind blows over the open water (called the fetch), the length of time the wind blows, the speed of the wind and the water depth.

Waves transport energy from where they were created by storms far out in the ocean to a shoreline. But a typical ocean wave does not resemble a perfect sinusoid, they are more irregular and complex than a simple sinusoidal wave. Only the steady up-and-down movement of a heavy swell resembles a sinusoidal wave much more than the chaotic nature of locally generated wind waves, as real sea waves contain a mixture of waves with different frequencies, wave heights and directions.

Wave Power Devices

Ocean wave energy has many advantages over ocean wind energy in that it is more predictable, less variable and offers higher available energy densities. Depending on the distance between the energy conversion device and the shoreline, wave energy systems can be classified as being either *Shoreline devices*, *Nearshore devices* or *Offshore devices*. So what is the difference between these three types of energy extraction devices.

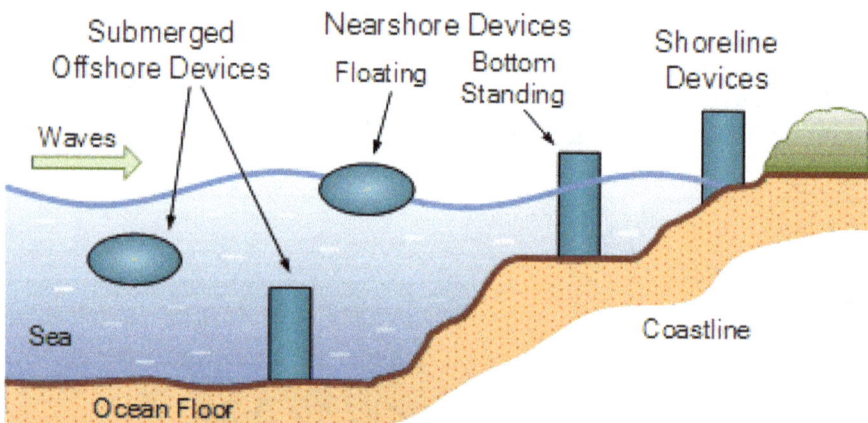

Shoreline devices are wave energy devices which are fixed to or embedded in the shoreline, that is they are both in and out of the water. Nearshore devices are characterised by being used to extract the wave power directly from the breaker zone and the waters immediately beyond the breaker zone, (i.e. at 20m water depth).

Offshore devices or deep water devices are the farthest out to sea and extend beyond the breaker lines utilising the high-energy densities and higher power wave profiles available in the deep water waves and surges.

One of the advantages of offshore devices is that there is no need for significant coastal earthworks, as there is with onshore devices.

As most of the energy within a wave is contained near the surface and falls off sharply with depth. There is a surprising range of designs available that maximise the energy available for capture. These wave energy devices are either fixed bottom standing designs used in shallow water and which pierce the water surface, or fully floating devices that are used to capture the kinetic energy content of a waves movement and convert each movement into electricity using a generator.

There are currently four basic "capture" methods:

- Point Absorbers: These are small vertical devices either fixed directly to the ocean floor or tethered via a chain that absorb the waves energy from all directions. These devices generate electricity from the bobbing or pitching action of a floating device. Typical wave energy devices include, floating buoys, floating bags, ducks, and articulated rafts, etc. These devices convert the up-and-down pitching motion of the waves into rotary movements, or oscillatory movements in a variety of devices to generate electricity. One of the advantages of floating devices over fixed devices it that they can be deployed in deeper water, where the wave energy is greater.

- Wave Attenuators: These are also known as "linear absorbers", are long horizontal semi-submerged snake-like devices that are oriented parallel to the direction of the waves. A wave attenuator is composed of a series of cylindrical sections linked together by flexible hinged joints that allow these individual sections to rotate and yaw relative to each other. The wave-induced motion of the device is used to pressurise a hydraulic piston, called a ram, which forces high pressure oil through smoothing accumulators to turn a hydraulic turbine generator producing electricity. Then wave attenuators convert the oscillating movement of a wave into hydraulic pressure.

- Oscillating Water Column: It is a partly submerged chamber fixed directly at the shoreline which converts wave energy into air pressure. The structure could be a natural cave with a blow hole or a manmade chamber or duct with an wind turbine generator located at the top well above the waters surface. The structure is built perpendicular to the waves so that the ebbing and flowing motion of the waves force the trapped water inside the chamber to oscillate in the vertical direction.

As the waves enter and exit the chamber, the water column moves up and down and acts like a piston on the air above the surface of the water, pushing it back and forth. This air is compressed and decompressed by this movement and is channelled through a wind turbine generator to produce electricity. The speed of air in the duct can be enhanced by making the cross-sectional area of the duct much less than that of the column.

- Overtopping Devices: It is also known as "spill-over" devices, are either fixed or floating structures that use ramps and tapered sides positioned perpendicular to the waves. The sea waves are driven up the ramp and over the sides filling-up a small tidal reservoir which is located 2 to 3 metres above sea level. The potential energy of the water trapped inside the reservoir is then extracted by returning the water back to the sea through a low head Kaplan turbine generator to produce electricity.

Then overtopping devices convert the potential energy available in the head of water into mechanical energy. The disadvantage of onshore overtopping schemes is that they have a relatively low power output and are only suitable for sites where there is a deep water shoreline and a low tidal range of less than about a metre.

The idea of harnessing the tremendous power of the oceans waves is not new. Like other forms of hydro power, wave energy does not require the burning of fossil fuels, which can pollute the air, contributing to acid rain and global warming. The energy is entirely clean and endlessly renewable. Wave power has many advantages compared to other forms of renewable energy with its main advantage being that it is predictable.

However, like many other forms of renewable energy, ocean wave energy also has its disadvantages such as its inflexible generation times dependant upon the tides, the visual impact of wave devices on the seas surface, as well as the threat of collision to shipping and navigation.

Here are some of the main advantages and disadvantages of wave energy:

Wave Energy Advantages

- Wave energy is an abundant and renewable energy resource as the waves are generated by the wind.

- Pollution free as wave energy generates little or no pollution to the environment compared to other green energies.

- Reduces dependency on fossil fuels as wave energy consumes no fossil fuels during operation.

- Wave energy is relatively consistent and predictable as waves can be accurately forecast several days in advance.

- Wave energy devices are modular and easily sited with additional wave energy devices added as needed.

- Dissipates the waves energy protecting the shoreline from coastal erosion.

- Presents no barriers or difficulty to migrating fish and aquatic animals.

Wave Energy Disadvantages

- Visual impact of wave energy conversion devices on the shoreline and offshore floating buoys or platforms.

- Wave energy conversion devices are location dependent requiring suitable sites were the waves are consistently strong.

- Intermittent power generation as the waves come in intervals does not generate power during calm periods.

- Offshore wave energy devices can be a threat to navigation that cannot see or detect them by radar.

- High power distribution costs to send the generated power from offshore devices to the land using long underwater cables.

- They must be able to withstand forces of nature resulting in high capital, construction and maintenance costs.

Run of River Hydroelectricity

Figure: Run-of-the-river systems like the one shown above tend to have larger hydroelectric flow rates than hydro dams that use reservoirs.

Run-of-the-river hydroelectric systems are hydroelectric systems that harvest the energy from flowing water to generate electricity in the absence of a large dam and reservoir—which is how they differ from conventional impoundment hydroelectric facilities. A small dam may be used to ensure enough water goes in the penstock, and possibly some storage (for same day use) The primary difference between this type of hydroelectric generation compared to others is that run-of-the-river primarily uses the natural flow rate of water to generate power—instead of the power of water falling a large distance. However, water may still experience some vertical drop in a run-of-the-river system from the natural landscape or small dam Another main difference between traditional hydropower is that run-of-the-river hydro is used in areas where there is little to no water storage, such as in a river.

There are several classifications of run-of-the-river systems, based primarily on their capacity. The types are outlined in the table below:

Classification	Capacity
Micro	< 100 kW
Mini	100 kW - 1 MW
Small	1 - 50 MW

It is important to note that some larger scale run-of-the-river plants exist, with outputs of hundreds or thousands of MW.

System Configurations

Run-of-river can mean several different things and have various configurations, as well as use different turbine technologies. Systems can involve the far upstream diversion of the river, diversion at an existing weir or dam, or simply in-stream current flow technology, and can involve penstock pipes, open channels, barrages and other diversion methods.

Penstock Based Systems or Streaming Systems

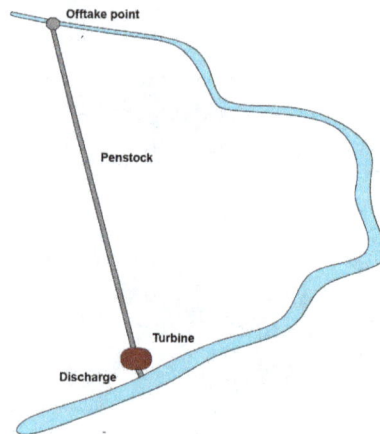

Figure: Penstock or streaming ROR system

This diversion is known as the "depleted reach" and is measured as the distance between where the water leaves a river to enter a hydro system and where the water goes back into the river. Taking a given volume of water out of a river will have an impact upon the natural ecology of the river bed. A large depleted reach may therefore compromise the ecology of the river and its environment. The smaller the depleted distance, the less impact there will be upon the fish and wildlife ecology.

Diversion Systems

In these systems a short open canal or closed pipe is used to divert water from up-stream of a low head artificial dam or natural drop, such as a waterfall or weir, into the turbine intake (Figure a). Typical examples are Inga 3 (Figure b), where a canal was constructed to divert the water flow upstream of the generator plant.

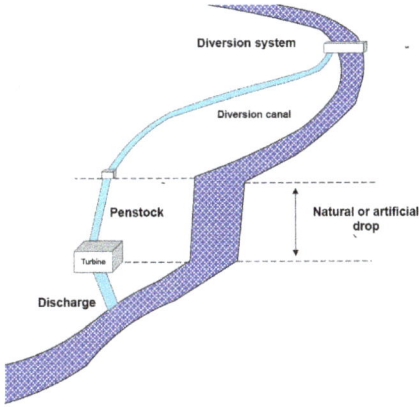

Figure a: River diversion system

Figure b: Inga 3 hydropower system

The drop across the Inga rapids or waterfall is approximately 96 m, although it is clear from figure b that the head created by the "dam" at Inga 3 is less than this. There is generally no limit on the amount of water which may be diverted. In many cases the turbine structure is incorporated into what amounts to a high weir or dam wall, the sole purpose of which is to feed water into the turbines.

Existing Dams and Weirs

ROR systems installed on existing dams use a diversion penstock around the wall of an existing dam with a low head, and significant inflow and outflow. Costs limit this approach to medium sized projects, i.e. the project must generate enough electricity to recover costs.

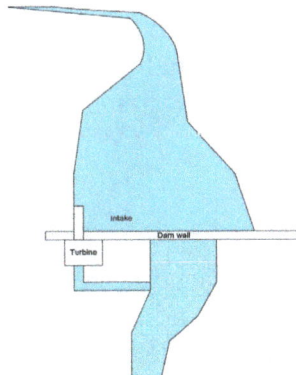

Figure: ROR system using existing dam

Water is returned to the river. The capacity of the system will depend on the existing flow of water through the dam required to maintain the river flow. Diversion of water for the ROR scheme may replace the existing water release from the dam.

Barrage Systems

In this case a barrage is constructed across the river forming a low head. The turbines, in-line reaction impeller types such as Kaplan turbines, are mounted either in horizontal or vertical configuration in the barrage and are driven by the flow of the river water through the barrage. This type of system has been used extensively on large, wide rivers with a constant flow rate.

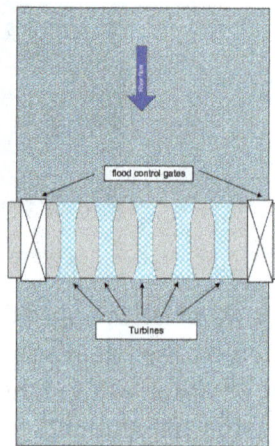

Figure: Barrage system

Instream Systems (River Current Systems)

Limited at the moment to smaller systems, these use reaction turbines or small bulb turbines installed in the body of the river and extract energy from the river current, being a truly ROR system. There are several different types of turbine being used, from fully immersed turbines to vertical shaft turbines where the generator is not immersed. There is considerable interest in the development of these systems as they involve minimal construction of additional infrastructure and do not interfere with the flow of water in any way. Instream systems can make use of existing structures such as bridges.

ROR System Turbines

Turbines used with ROR systems may be impulse type or reaction type, and depend on the size of the system installed. Kaplan and Francis turbines are commonly used with kinetic systems, while Pelton wheel types are usually found in systems relying on a substantial head of water, such as the streaming system.

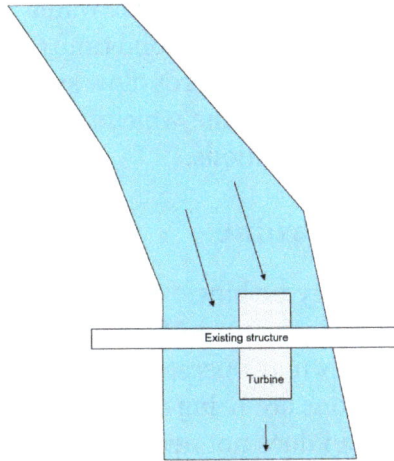

Figure: Instream systems

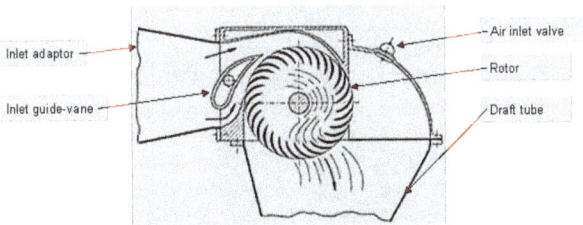

Figure: Vertical crossflow turbine

Recent Developments in Turbines using ROR Techniques

Virtually a modern version of the waterwheel, the VCT is well adapted to ROR operation. Available in sizes ranging from tens of kW to several MW, this unit is essentially a micro-turbine, but could be applied easily to many of the smaller dams and water structures in the country. These turbines do not suffer damage from small debris in the water and are fairly rugged. The crossflow principle can be applied in either a horizontal inflow or a vertical inflow method.

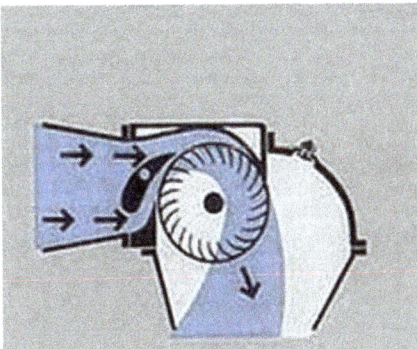

Figure a: Crossflow inflow horizontal

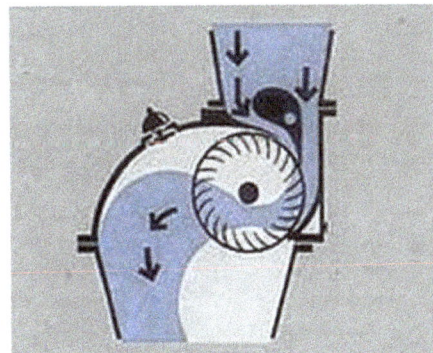

Figure b: Crossflow inflow vertical

The main feature of the crossflow turbine is that water passes through the turbine blades twice. The first time on entry and the second time on exit after travelling through the centre of the blade wheel. Dual vertical crossflow systems, consisting of two cross-flow turbines mounted vertically in a frame which spans the width of the feed canal have been used in some smaller installations.

Hydrofoil Kinetic Energy Turbine

A truly ROR turbine, kinetic energy turbines, also called free-flow turbines, generate electricity from the kinetic energy present in flowing water rather than the potential energy from the head. The systems may operate in rivers or man-made channels. Both vertical and horizontal axis versions are being developed. The vertical axis turbine can be placed in the river directly, and does not need a separate structure to control water flow. The vertical axis turbine has the advantage that the generator can be mounted above the turbine and therefore out of the water.

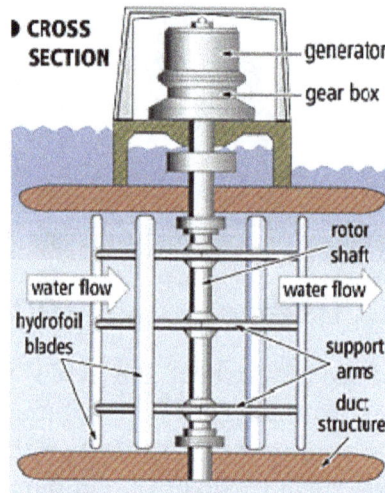

Figure a: Vertical axis turbine (Blue Energy Systems).

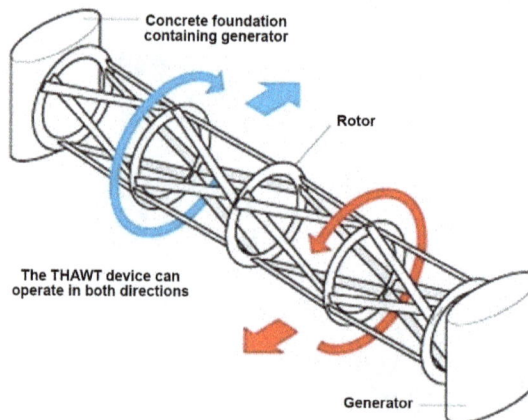

Figure b: The transverse horizontal axis water turbine

Horizontal Axis "Bulb" Turbine

This is a more modern development using a Kaplan turbine for energy capture, the generator and control circuits are contained in a bulb and the whole device installed in a cone or cylinder to control water flow.

Archimedean Screw Turbines

These systems work on the principle of the Archimedean screw pump in reverse. . Sizes from 1 kW to 150 kW capacity are in service. The systems are fish friendly and are not susceptible to blocking by small debris. This is a true run-of-river device as the in feed takes place at the water surface, and the speed of rotation is dependant on the speed of the water entering the screw. The screw can be enclosed, but is in many cases left open.

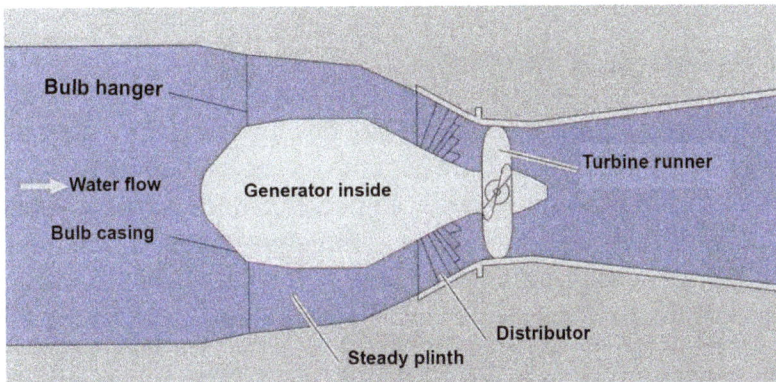

Figure: Bulb Turbine

Power Developed by ROR Systems

There are two methods which can be used for calculating the power generated by run-of-river systems, the first based on the potential energy in stored water or head of water, and the second based on the kinetic energy of flowing water.

Potential Energy Method

Potential energy (PE) per unit volume.

$$PE = \rho gh$$

Where:

ρ = the density of the water (10^3 kg/m³),

h = the head of water (m)

g = the gravitational constant (9,81 m/s²)

The power P_{max} of the hydro system (in kW) is given by,

$$P_{max} = \eta \rho g h Q$$

Where:

Q is the volume of water flowing per second (the flow rate in m³/s) η is the efficiency of the turbine,

For water flowing at 1 m³/s from a head of 1 m, the power generated is equivalent to 9,81 kW assuming an energy conversion efficiency of 100% or just over 9 kW with a turbine efficiency of between 90% and 95%.

This simple looking formula usually applies to static hydro, using stored water. If applied to dynamic or kinetic systems, some attention needs to be given to the calculation of the values and how they influence the power output.

Kinetic Energy Method

The turbine converts the kinetic energy of the flowing water into the rotational (mechanical) energy of the turbine and the generator. The available energy therefore depends on the quantity of water flowing through the turbine and the square of its velocity. Impulse turbines which are only partially submerged are more commonly employed in fast flowing run-of-river installations while In deeper, slower flowing rivers, submerged Kaplan turbines may be used to extract the energy from the water flow.

The maximum power output from a turbine used in a kinetic ROR application is proportional to the kinetic energy of the water flowing through the turbine . Taking the efficiency η of the turbine into account, the maximum output power P_{max} is given by:

$$P_{max} = \tfrac{1}{2}\, \eta \rho Q v^2$$

Where:

v = the velocity of the water flow

Q = the volume of water flowing through the turbine per second

ρ = the density of water (10^3 kg/m³)

but $Q = Av$, where A is the swept area of the turbine blades.

Thus,

$$P_{max} = \tfrac{1}{2}\eta \rho A v^3$$

This relationship is directly analogous to the equation for the theoretical power generated by wind turbines. Note that the power output is proportional to the cube of the velocity of the water.

If we return to equation $P_{max} = \eta\rho ghQ$ and use the Bernoulli equation,

$$v = \sqrt{2gh} \; or \; v^2 \; 2gh$$

to get,

$$h = \frac{v2}{2g}$$

Substituting for h in equation $P_{max} = \frac{1}{2}\eta\rho Av^3$.

Gives,

$$P_{max} = \frac{1}{2}\eta\rho v^2 Q$$

Now $Q = vA$ so $P_{max} = \frac{1}{2}\eta\rho Av^3$, the same result as in equation $P_{max} = \frac{1}{2}\eta\rho Av^3$.

The result is:

- If the head or pressure drop is known but not the velocity, use equation $P_{max} = \eta\rho ghQ$

- If the velocity is known neglect the head and use equation $P_{max} = \frac{1}{2}\eta\rho Av^3$

Thus the power generated by one cubic metre of water flowing at one metre per second through a turbine with 100% efficiency will be 0,5 kW or slightly less taking into account the inefficacies in the system. This is only one twentieth of the power generated by the same volume flow from the static system in the example above. To generate the same power with the same volume of water from a run-of-river installation, the speed of the water flow should be 3√20 metres per second (2,7 m/s).

Comparison to Traditional Hydro

There are several benefits that exist as a result of using run-of-the-river hydro instead of traditional, dam based hydro. First, traditional hydro dams are expensive and time consuming to build. Comparatively speaking, run-of-the-river systems are less expensive to build and can be built over a shorter period of time. In addition, many areas where large hydro is used frequently—such as in Canada—have developed many of the existing favourable hydropower sites.Run of river systems also avoid some of the environmental problems associated with the flooding, since the pondage is much smaller than the lakes for traditional hydro.

Although there are some favourable aspects to run-of-the-river hydro, the output is significantly lower than large scale hydro. Dam based hydroelectric generation usually provides a lower cost per kWh despite the larger initial investment. Therefore, the absence of the large dam and reservoir means that the power plant will be less reliable for electricity generation. If the water levels are depleted upstream, potentially due to drought conditions, there is less water to run the hydroelectric system.

Environmental Impact

Compared to the combustion of fossil fuels, run-of-the-river hydro is responsible for fewer greenhouse gas emissions. Most of these emissions are a result of the construction of the system itself, but the operation of the plant itself contributes almost no emissions. Run-of-the-river hydroelectric plants can also be beneficial compared to impoundment dams as the small amount of water storage results in a smaller environmental footprint when compared to dams with large amounts of water storage.

Although the emission impact of these systems is less than the combustion of fuels, there are other environmental impacts that must be taken into account. First, the manipulation of river flows can cause a significant number of environmental impacts. Any diversion in the river changes how the aquatic ecosystem works, which could affect fish populations and the health of the river overall. However, the small head difference in run-of-the-river systems allows for the construction of fish ladders, which could allow fish to move through the system without being harmed or having their migration routes interrupted. As well, the changes in the river basin or water composition could increase species mortality, disrupt migration, or cause an imbalance in biodiversity. Finally, thermal pollution and increased turbidity of the exiting water are possible side effects of sending water through turbines and back into a river.

Overall, it is difficult to determine in general whether or not the damage inflicted on the environment by run-of-the-river systems is outweighed by the relatively small output as compared to large hydro dams. That means that each project must be evaluated on the specific details of the hydroelectric power plants being proposed.

Advantages

When developed with care to footprint size and location, ROR hydro projects can create sustainable energy minimizing impacts to the surrounding environment and nearby communities. Advantages include:

Cleaner Power, Fewer Greenhouse Gases

Like all hydro-electric power, run-of-the-river hydro harnesses the natural potential energy of water, eliminating the need to burn coal or natural gas to generate the electricity needed by consumers and industry. Moreover, run-of-the-river hydro-electric plants do not have reservoirs thus eliminating the methane and carbon dioxide emissions caused by the decomposition of organic matter in the reservoir of a conventional hydro-electric dam. This is a particular advantage in tropical countries where methane generation can be a problem.

Less Flooding/Reservoirs

Without a reservoir, flooding of the upper part of the river does not take place. As a result, people remain living at or near the river and existing habitats are not flooded. Any pre-existing pattern of flooding will continue unaltered, presenting a flood risk to the facility and downstream areas.

Disadvantages

Unfirm Power

Run-of-the-River power is considered an "unfirm" source of power: a run-of-the-river project has little or no capacity for energy storage and hence can't co-ordinate the output of electricity generation to match consumer demand. It thus generates much more power during times when seasonal river flows are high (i.e., spring freshet), and depending on location, much less during drier summer months or frozen winter months.

Availability of Sites

Rapids can provide enough hydraulic head

The potential power at a site is a result of the head and flow of water. By damming a river, the head is available to generate power at the face of the dam. Where a dam may create a reservoir hundreds of kilometres long, in run of the river the head is usually delivered by a canal, pipe or tunnel constructed upstream of the power house. Due to the cost of upstream construction, a steep drop is desirable, such as falls or rapids.

Environmental Impacts

Small, well-sited ROR projects can be developed with minimal environmental impacts. Larger projects have more environmental concerns. In the case of fish-bearing rivers a ladder may be required and dissolved gases downstream may affect fish.

In British Columbia the mountainous terrain and wealth of big rivers have made it a global testing ground for 10-50Mw run-of-river technology. As of March 2010, there were 628 applications pending for new water licences solely for the purposes of power generation – representing more than 750 potential points of river diversion.

Concerns

- Diverting large amounts of river water reduces river flows, affecting water velocity and depth, minimizing habitat quality for fish and aquatic organisms; reduced flows can lead to excessively warm water for salmon and other fish in summer.

- In undeveloped areas, new access roads and transmission lines can cause habitat fragmentation, allowing the introduction of invasive species.

- The lack of reservoir storage may result in intermittent operation, reducing the project's viability.

Pumped-storage Hydroelectricity

This method produces electricity to supply high peak demands by moving water between reservoirs at different elevations. At times of low electrical demand, the excess generation capacity is used to pump water into the higher reservoir. When the demand becomes greater, water is released back into the lower reservoir through a turbine. Pumped-storage schemes currently provide the most commercially important means of large-scale grid energy storage and improve the daily capacity factor of the generation system.

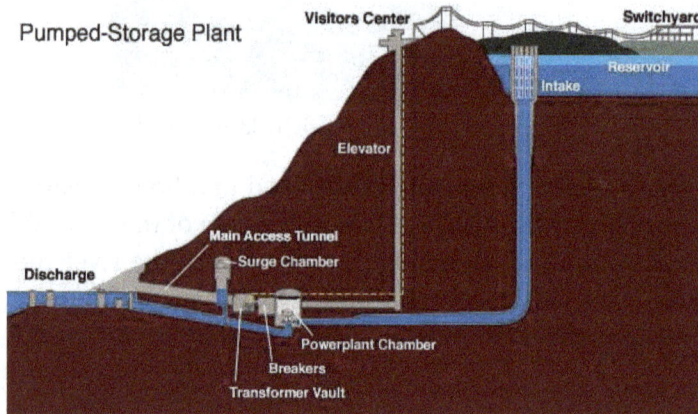

Figure: Diagram of the TVA pumped storage facility at Raccoon Mountain Pumped-Storage Plant.

Pumped storage is the largest-capacity form of grid energy storage available, The main disadvantage of PHS is the specialist nature of the site required, needing both geographical height and water availability. Suitable sites are therefore likely to be in hilly or mountainous regions, and potentially in areas of outstanding natural beauty, and therefore there are also social and ecological issues to overcome.

Figure: Pumped-storage hydroelectricity – the upper reservoir (Llyn Stwlan) and dam of the Ffestiniog Pumped Storage Scheme in north Wales. The lower power station has four water turbines which generate 360 MW of electricity within 60 seconds of the need arising

At times of low electrical demand, excess generation capacity is used to pump water into the higher reservoir. When there is higher demand, water is released back into the lower reservoir through a turbine, generating electricity. Reversible turbine/generator assemblies act as pump and turbine (usually a Francis turbine design). Nearly all facilities use the height difference between two natural bodies of water or artificial reservoirs. Pure pumped-storage plants just shift the water between reservoirs, while the "pump-back" approach is a combination of pumped storage and conventional hydroelectric plants that use natural stream-flow. Plants that do not use pumped-storage are referred to as conventional hydroelectric plants; conventional hydroelectric plants that have significant storage capacity may be able to play a similar role in the electrical grid as pumped storage, by deferring output until needed.

Taking into account evaporation losses from the exposed water surface and conversion losses, energy recovery of 80% or more can be regained. The technique is currently the most cost-effective means of storing large amounts of electrical energy on an operating basis, but capital costs and the presence of appropriate geography are critical decision factors.

The relatively low energy density of pumped storage systems requires either a very large body of water or a large variation in height. For example, 1000 kilograms of water (1 cubic meter) at the top of a 100 meter tower has a potential energy of about 0.272 kW·h (capable of raising the temperature of the same amount of water by only 0.23 Celsius = 0.42 Fahrenheit). The only way to store a significant amount of energy is by having a large body of water located on a hill relatively near, but as high as possible above, a second body of water. In some places this occurs naturally, in others one or both bodies of water have been man-made. Projects in which both reservoirs are artificial and in which no natural waterways are involved are commonly referred to as "closed loop".

This system may be economical because it flattens out load variations on the power grid, permitting thermal power stations such as coal-fired plants and nuclear power plants that provide base-load electricity to continue operating at peak efficiency (Base load power plants), while reducing the need for "peaking" power plants that use the same fuels as many base load thermal plants, gas and oil, but have been designed for flexibility rather than maximal thermal efficiency. However, capital costs for purpose-built hydro-storage are relatively high.

Along with energy management, pumped storage systems help control electrical network frequency and provide reserve generation. Thermal plants are much less able to respond to sudden changes in electrical demand, potentially causing frequency and voltage instability. Pumped storage plants, like other hydroelectric plants, can respond to load changes within seconds.

The important use for pumped storage is to level the fluctuating output of intermittent energy sources. The pumped storage provides a load at times of high electricity output and low electricity demand, enabling additional system peak capacity. In certain jurisdictions, electricity prices may be close to zero or occasionally negative (Ontario in early September, 2006), on occasions that there is more electrical generation than load available to absorb it; although at present this is rarely due to wind alone, increased wind generation may increase the likelihood of such occurrences. It is particularly likely that pumped storage will become especially important as a balance for very large scale photovoltaic generation.

Potential Technologies

Seawater

Pumped storage plants can operate with seawater, although there are additional challenges compared to using fresh water. In 1999, the 30 MW Yanbaru project in Okinawa was the first demonstration of seawater pumped storage. It has since been decommissioned. A 300 MW seawater-based Lanai Pumped Storage Project was considered for Lanai, Hawaii, and seawater-based projects have been proposed in Ireland. A pair of proposed projects in the Atacama Desert in northern Chile would use 600 MW of photovoltaic solar (Skies of Tarapacá) together with 300 MW of pumped storage (Mirror of Tarapacá) raising seawater 600 metres (2,000 ft) up a coastal cliff.

Underground Reservoirs

The use of underground reservoirs has been investigated. Recent examples include the proposed Summit project in Norton, Ohio, the proposed Maysville project in Kentucky (underground limestone mine), and the Mount Hope project in New Jersey, which was to have used a former iron mine as the lower reservoir. The proposed energy storage at the Callio site in Pyhäjärvi (Finland) would utilize the deepest base metal mine in Europe, with 1,450 metres (4,760 ft) elevation difference. Several new underground pumped storage projects have been proposed. Cost-per-kilowatt estimates for these projects can be lower than for surface projects if they use existing underground mine space. There are limited opportunities involving suitable underground space, but the number of underground pumped storage opportunities may increase if abandoned coal mines prove suitable.

In Bendigo, Victoria, Australia, the Bendigo Sustainability Group has proposed the use of the old gold mines under Bendigo for Pumped Hydro Energy Storage. Bendigo has the greatest concentration of deep shaft hard rock mines anywhere in the world with over 5,000 shafts sunk under Bendigo in the second half of the 19th Century. The deepest shaft extends 1,406 metres vertically underground. A recent pre-feasibility study has shown the concept to be viable with a generation capacity of 30 MW and a run time of 6 hours using a water head of over 750 metres.

Decentralised Systems

Small pumped-storage hydropower plants can be built on streams and within infrastructures, such as drinking water networks and artificial snow making infrastructures. Such plants provide distributed energy storage and distributed flexible electricity production and can contribute to the decentralized integration of intermittent renewable energy technologies, such as wind power and solar power. Reservoirs that can be used for small pumped-storage hydropower plants could include natural or artificial lakes, reservoirs within other structures such as irrigation, or unused portions of mines or underground military installations. In Switzerland one study suggested that the total installed capacity of small pumped-storage hydropower plants in 2011 could be increased by 3 to 9 times by providing adequate policy instruments.

Underwater Reservoirs

In March 2017 the research project StEnSea (Storing Energy at Sea) announced their successful completion of a four-week test of a pumped storage underwater reservoir. In this configuration a hollow sphere submerged and anchored at great depth acts as the lower reservoir, while the upper reservoir is the enclosing body of water. Electricity is created when water is let in via a reversible turbine integrated into the sphere. During off-peak hours the turbine changes direction and pumps the water out again, using "surplus" electricity from the grid. The quantity of power created when water is let in grows proportionally to the height of the column of water above the sphere, in other words: the deeper the sphere is located, the more potential energy it can store, which can be transformed into electric power. On the other hand, pumping the water back out at greater depths also uses up more power, since the turbine-turned-pump must act on the same entire column of water.

As such the energy storage capacity of the submerged reservoir is not governed by the gravitational energy in the traditional sense, but rather by the vertical pressure variation.

While StEnSea's test took place at a depth of 100 m in the fresh water Lake Constance, the technology is foreseen to be used in salt water at greater depths. Since the submerged reservoir needs only a connecting electrical cable, the depth at which it can be employed is limited only by the depth at which the turbine can function, currently limited to 700 m. The challenge of designing salt water pumped storage in this underwater configuration brings a range of advantages:

- No land area is required,

- No mechanical structure other than the electrical cable needs to span the distance of the potential energy difference,

- In the presence of sufficient seabed area multiple reservoirs can scale the storage capacity without limits,

- Should a reservoir collapse, the consequences would be limited apart from the loss of the reservoir itself,

- Evaporation from the upper reservoir has no effect on the energy conversion efficiency,

- Transmission of electricity between the reservoir and the grid can be established from a nearby offshore wind farm limiting transmission loss and obviating the need for onshore cabling permits.

A current commercial design featuring a sphere with an inner diameter of 30 m submerged to 700 m would correspond to a 20 MWh capacity which with a 5 MW turbine would lead to a 4-hour discharge time. An energy park with multiple such reservoirs would bring the storage cost to around a few eurocents per kWh with construction and equipment costs in the range €1,200-€1,400 per kW. To avoid excessive transmission cost and loss, the reservoirs should be placed off deep water coasts of densely populated areas, such as Norway, Spain, USA and Japan. With this limitation the concept would allow for worldwide electricity storage of close to 900 GWh.

For comparison, a traditional, gravity-based pumped storage capable of storing 20 MWh in a water reservoir the size of a 30 m sphere would need a hydraulic head of 519 m with the elevation spanned by a pressurized water pipe requiring typically a hill or mountain for support.

Home use

Using a pumped-storage system of cisterns and small generators, pico hydro may also be effective for "closed loop" home energy generation systems.

Pico Hydropower

Pico hydro is hydro power with a maximum electrical output of five kilowatts (5kW). Hydro power systems of this size benefit in terms of cost and simplicity from different approaches in the design, planning and installation than those which are applied to larger hydro power. Recent innovations in Pico-hydro technology have made it an economic source of power even in many of the world's poorest and most inaccessible areas. It is also a versatile power source. AC electricity can be produced enabling standard electrical appliances to be used. Common examples of devices which can be powered by Pico-hydro are light bulbs, radio and televisions. Normally, Pico-hydropower system is found at rural or hilly area. Figure below shows an example of typical Pico hydro system applications at hilly area. This system will operate using upper water reservoir which is a few meter high from ground. From the reservoir, water flows downhill through the piping system. This downhill distance is called "head" and it allows the water to accelerate for prime moving system. Thus, the turbine will rotate the alternator to produce electricity. However, this research is conducted to show the potential of consuming water distributed to houses at town area as an alternative of renewable energy source.

The water flow inside the pipelines has potential of kinetic energy to spin small scale generator turbine for electricity generation. Therefore, this project has been done to show the additional use of consuming water distributed to houses for electrical power generation instead of routine activities such as bathe, laundry and dish wash. The electricity can be generated at the same time those usual activities are done without extra charge on the water bill consumption. The main function of the system is to store the generated power by means of battery charging for future use particularly during electricity blackouts. The proposed system is expected has a maximum capacity of 10W which is very much less compared to other Pico-hydro power systems.

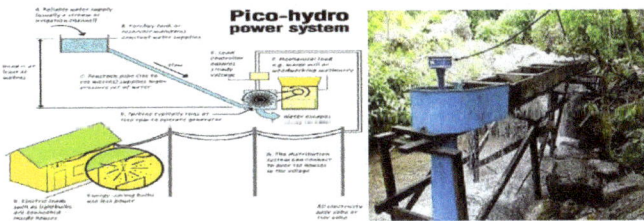

Figure: Pico-hydro power system applications at rural area

Pico-hydro System Planning & Development

There are many factors that determine the feasibility and achievability of the system. This includes:

i.   The amount of power available from the water flow inside the pipelines. This depends on the water pressure, amount of water available and friction losses in the pipelines.

ii. The turbine type and availability of required generator type and capacity.

iii. The types and capacity of electrical loads to be supplied by the Pico-hydro system.

iv. The cost of developing the project and operating the system.

(A) Power Estimation: In general, the feasibility of the proposed Pico-hydro system is based on the following potential input and output power equation:

$$P_{in} = H \times Q \times g \quad (1)$$

$$P_{out} = H \times Q \times g \times \eta \quad (2)$$

Where,

P_{in} = Input power (Hydro power)

P_{out} = Output power (Generator output)

H = Head (meter)

Q = Water flow rate (liter/second)

g = gravity (9.81 m/s2)

η = efficiency

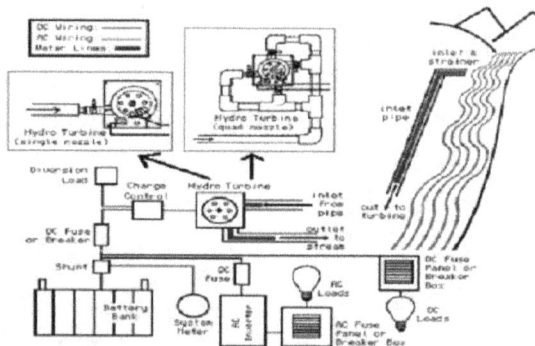

Figure: Probable Pico-hydro system: Sustainable Renewable Energy Solution

Based on the equation ($P_{in} = H \times Q \times g$ (1)) and ($P_{out} = H \times Q \times g \times \eta$ (2)), both head and water flow rate are very important parameters in hydropower system. Head is a measure of falling water at turbine, i.e. vertical distance from the top of the penstock to the turbine at the bottom. Conversely, water flow rate is the amount of water flows within one second. Normally, water flow available is more than needed since the flows for Pico-hydro are small. Thus, it is important to measure the head carefully because the greater head, the greater power and the higher speed of the turbine rotation.

(B) Head Measurement: When determining head (falling water), gross or "static" head and net or "dynamic" head must be considered. Gross head is the vertical distance between the top of the penstock and the point where the water hits the turbine. Net head is gross head minus the pressure or head losses due to friction and turbulence in the penstock. These head losses depend on the type, diameter, and length of the penstock piping, and the number of bends or elbows. Gross head can be used to estimate power availability and determine general feasibility, but net head is used to calculate the actual power available. There are many methods of head measurement. However, since the proposed Pico-hydro system uses consuming water distributed to houses supplied by the Water Utility Company whereby the utility water tank can be very far from the houses, the simplest and most practical method for head measurement is water-filled tube and calibrated pressure gauge.

Through this method, the pressure gauge reading in psi can be converted to head in meters using the following equation of pressure to head conversion:

$$H = 0.704 \times P \ (3)$$

Where = Head (meter) & P = Pressure (psi)

Equation above shows that the water pressure at consumers" end is a very important parameter to be determined in the design and development of the proposed Pico-hydro system. The water pressure represents the net head of the system that useful to calculate the actual power available.

(C) Water Flow Rate Measurement: The most simple of flow measurement for small streams is the bucket method. Therefore, this method has been used due to the capacity of the proposed hydropower system is significantly small. Moreover, this method is considerably practical due to the proposed hydropower system is very uncommon compared to other system in its category in which the source of energy is from the consuming water distributed to houses by the Water Utility Company. Throughout this method, the flow rate of the distributed water is diverted into a bucket or barrel and the time it takes for the container to fill is recorded. The volume of the container is known and the flow rate is simply obtained by dividing this volume by the filling time. For example, the flow rate of water that filled 20 litres bucket within one minute is 20 litres per minute or 0.333 l/s. This is repeated several times to give more consistent and accurate measurement.

(D) Pipeline System and Friction Loss: Piping system is used to carry water to a turbine. This is commonly termed as penstock which consists of pipe from the reservoir or fore bay tank to the turbine and valve or gate that controls the rate of water flow. The proposed Pico-hydro system will have the water source from consuming water distributed to houses. Thus, the system must be designed with ability to produce high water pressure to rotate the turbine at the most possible speed and at the same time the water can be recycled and used to other routine activities such as bathe and laundry. Thus, no extra charge on the water bill consumption incurred. In order to do so, a suitable piping scheme with appropriate nozzle between the source (consumers" end) and the turbine is required to maximize the turbine rotation speed.

Figure: Pen stroke for Pico-hydro system

In real fluid flows, friction losses occur due to the resistance of the pipe walls and the fittings. This leads to an irreversible transformation of the flowing fluid energy into heat. Friction head loss is divided into two main categories, "major losses" associated with energy loss per length of pipe, and "minor losses" associated with bends, fittings, valves, etc. For the proposed system, the hydro power available at the consumers" water outlet (consumers" end) is the net power after taking into account the friction losses along the pipelines from the utility tank to the consumers. Thus, consideration on the pipe length and diameter to handle the amount of water flow and piping accessories to convey the water to the turbine is very important to minimize the friction loss for the piping scheme between the source and the turbine of the Pico-hydro power system. This can be done by appropriately select the diameters and types of bends, fittings and valve and minimizing the use of these accessories. Moreover, it is necessary to minimize the piping system length between the water source and the turbine although it is extremely short when compared to the main pipeline from the utility tank to consumers" end. By considering all these matters, the proposed piping Pico-hydro piping scheme is assumed to have minor friction losses or can be neglected. This means the net hydro power at consumers" end is more or less similar to hydro power to turbine. As a result, the following are the list of piping accessories that used in the system:

i.   Valve – ball valve

ii. Nozzle – variable

iii. Elbow – 90 degree

iv. Tee – flanged

v.   Straight connector

vi. Pressure Gauge – 0 to 10 bar

vii. Main pipe – diameter 1.5mm

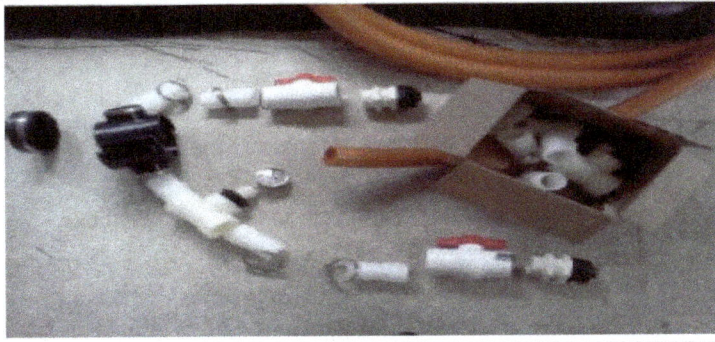

Figure: Pico hydro Piping Control elements

(E) Selection of Generator: Generating system for a hydro power scheme is selected based on the following concerns:

i.   The estimated power of a hydropower system.

ii. Type of supply system and electrical load: AC or DC.

iii. Available generating capacity in the market.

iv. Generator with cost effective.

Normally, Pico-hydro systems use AC generator either induction or synchronous machine type. This is because the system is used to supply AC electrical appliances and DC generator with size above 2kW is said expensive and has brush gear that requires appreciable maintenance. In addition, DC switches for the voltages and currents concerned are more expensive than their AC equivalents. However, in this project, a brush permanent magnet DC Generator is preferred as the main function of the proposed Pico-hydro system is for energy storage (battery charging). One significant advantage of using DC type of permanent magnet generator over AC generator is that DC generator is designed to provide high currents at minimum voltage requirement for the charging of battery and operation of direct current loads.

This is related with the load type to be supplied. Moreover, permanent magnet generator is selected as it is much cheaper and has smaller overall size rather than of wound field. Other than that, this type of generator is more efficient because no power is wasted to generate the magnetic field. Hence, permanent magnet DC generator manufactured by "Wind stream Power" with maximum power output of 80W at 2800rpm shaft rotation is chosen. The capacity of the generator is considerably high compared to the estimated power mentioned earlier, i.e. 10W. This is because it is very difficult to find the generating capacity available in the market that match with the estimated power. The maximum current at continuous duty is 1.5A whilst maximum current for 10minutes duty is limited to 2.5A. One indisputable problem that will be encountered when using this generator is high torque during the shaft.

Figure: Pico hydro Generator rotation.

This makes the rotation of the generator is not at the highest possible speed. Research on the generator design itself to maximize the system efficiency will be done in the future.

(F) Selection of Turbine Type: Selection of turbine to be used is very important in the design and development of a hydropower system. Table shows the groups of impulse and reaction turbines that are available. In general, reaction turbine is fully immersed in water and is enclosed in a pressure casing. The runner or rotating element and casing are carefully engineered so that the clearance between them is minimized. In contrast, impulse turbine can operate in air and works with high-speed jet of water.

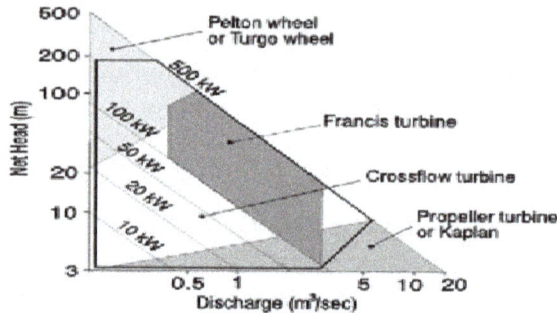

Figure: Head-flow ranges of small hydro turbines

Usually, impulse turbines are cheaper than reaction turbines because no specialist pressure casing and no carefully engineered clearance are needed.

Pelton Turbine: Pelton turbines are suitable to high head, low flow applications. It can be outfitted with one, two, or more nozzles for higher output. Typically, when using this type of turbine, water is piped down a hillside so that at the lower end of the pipe, it emerges from a narrow nozzle as a jet with very high velocity to the turbine blades.

The similar concept of water jet is also utilized in the proposed Pico-hydro system; instead the water is tapped from the domestic consuming water outlet. After a series of performance test on various constructions of Pelton turbine blades, the turbine with 6 cups fabricated from polyvinyl chloride (PVC) as shown in figure below is used for the proposed system.

The advantages of using PVC are less expensive and not require specialist to fabricate the turbine blades. Meanwhile, the base that associated with the drive shaft (rotor) for the turbine blades mounting is fabricated from aluminum with 80mm diameters. Small size of turbine is important to ensure the generator shaft to rotate at an optimum speed.

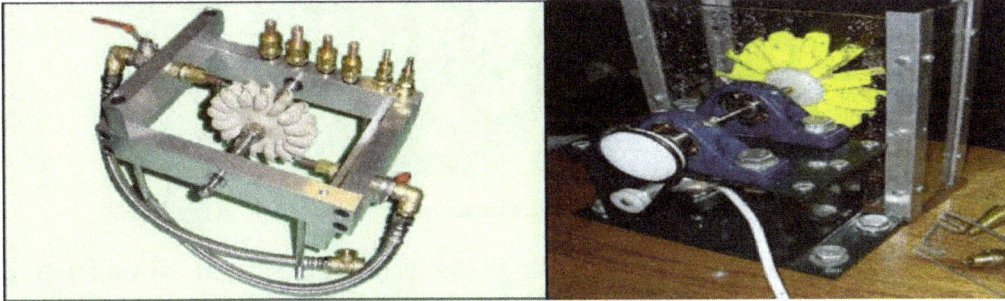

Figure: Turbine for Pico Hydro System

Table: Groups of Impulse and Reaction Turbine.

Turbine Runner	Head Pressure		
	High	Medium	Low
Impulse	Pelton Turgo Multi-jet Pelton	Cross flow Turgo Multi-jet Pelton	Cross flow
Reaction		Francis Pump-as-turbine	Propeller Kaplan

(G) Types of Electrical Loads: Electrical loads that are normally connected to a Pico-hydro system at rural area are lighting, battery chargers, radios, televisions, ventilation fans and refrigerators. For the proposed Pico-hydro system, however, the generating capacity is much lower compared to the existing Pico-hydro system at rural area. Thus, the main function of the proposed system is for battery charging. A battery allows the future use of small electrical loads and can be recharged when required. Examples of future use of small loads particularly during electricity blackouts are LED lighting, mobile phone battery charging and toys battery charging. There are two common types of rechargeable batteries used for providing power to small loads which are lead-acid and nickel-cadmium (Ni-Cad). Battery size depends on the generating capacity of the proposed Pico-hydro system and its application. For this type of Pico-hydro system, Ni-Cad battery is preferred and more practical as it is easier to handle and reliable. Moreover, Ni-Cad is the opposite to lead acid in that it performs better and last longer if fully discharged before re-charging. On the other hand, lead acid batteries should not be fully discharged as this damages them.

(H) Battery Charger: As illustrated in figure below, the generator output is connected to the charging circuits for energy storage purpose. Simple charging circuit as shown in figure below is used. The battery charger is suitable for 9V to 12V batteries. For charging purpose, the maximum load current is limited to 1.5A. This is based on the maximum load current of the LM317 voltage regulator and the maximum current at continuous duty of the generator.

Figure: Pico hydro in operation

In addition, due to the generating capacity of the Pico-hydro system, Ni-Cad battery is preferred. Figure above also illustrates that there is a point of direct output from generator. This is to offer other application that might be relevant.

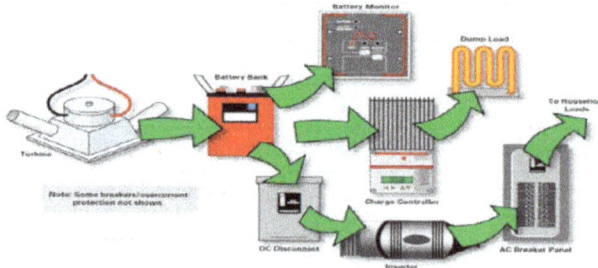

Figure: Operation of Pico hydro with modern controls

Micro Hydropower

Micro hydropower is the small-scale harnessing of energy from falling water, such as steep mountain rivers. Using this renewable, indigenous, non-polluting resource, micro-hydro plants can generate power for homes, hospitals, schools and workshops. The micro hydropower station, which converts the energy of flowing water into electricity, provides poor communities in rural areas with an affordable, easy to maintain and long-term solution to their energy needs.

A micro hydro power (MHP)'plant' is a type of hydro electric power scheme that produces up to 100 KW of electricity using a flowing steam or a water flow. The electricity from such systems is used to power up isolated homes or communities and is sometimes connected to the public grid.

Micro hydro systems are generally used in developing countries to provide electricity to isolated communities or rural villages where electricity grid is not available. Feeding back into the national grid when electricity production is in surplus is also evident in some cases. The micro hydro scheme design can be approached as per household basis or at the village level often involving local materials and labor.

In 1995, the micro-hydro capacity in the world was estimated at 28 GW, supplying about 115 TWh of electricity. About 60% of this capacity was in the developed world, with 40% in developing areas.

Micro hydro plants that are found in the developing world are mostly in mountainous regions for instance in the some places in the Himalayas as well as in Nepal where there are around 2,000 schemes, including both mechanical and electrical power generation. In South America, there are micro-hydro programs in the countries along the Andes, such as Peru and Bolivia. Smaller programs have also been set up in the hilly areas of Sri Lanka, Philippines and some parts of China.

Technology

Scheme Components

Most micro-hydro systems are 'run-of-river' which means that they don't need large dams to store water. However, they do need some water-management systems.

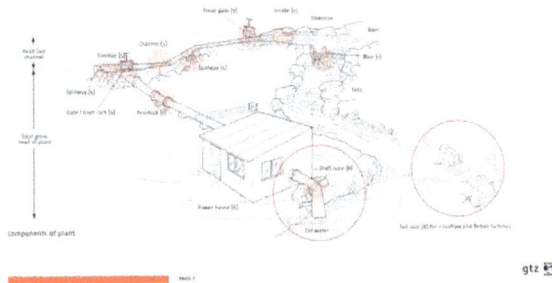

Components Micro Hydro Power Plant

The illustration above shows just how a micro-hydro system can be setup. For water diversion the river water level has to be raised by a barrier, the weir. The water is diverted at the intake and conveyed by the channel along the landscape´s contour lines. The spillways protect against damage from excessive water flow. Water is slowed down and collected in the fore bay, from where it enters into the penstock; the pressure pipe conveys the water to the power house where the power conversion turbine, mill or generating equipment is installed. The turbine is the core of a MHP, which is rotated by the moving water. Different types of turbines are used depending on the head and the flow of the site the turbines are used to rotate a shaft which is then used to drive the generator. The water is then discharged via the draft tube or a tail race channel in case of cross flow or Pelton turbines.

Due to the nature of the micro-hydro schemed to be remote; a local grid is constructed to distribute the electricity to the different users. The demand output must match the capacity of the generator otherwise the voltage and frequency can vary suddenly, which can result in the damage of certain electrical equipment. The power demand in an off-grid is often variable since people switch lights and machines on and off, so the supply from the micro-hydro system must be varied to keep close control. This can be done by varying the water flow, or by using an electronic load controller.

Suitable Conditions for Micro Hydro Power Plants

The ideal geographical areas for exploiting small scale hydro schemes is where there are steep rivers flowing all year round. Islands with moist marine climates are also suitable. Low-head turbines have been developed for small-scale exploitation of rivers or irrigation canals where there is a small head but sufficient flow to provide adequate power.

To understand more about a suitable potential site, the hydrology of the site needs to be known and a site survey carried out so as to determine the actual flow and head data. Hydrological information is easily accessible from the metrological or irrigation department of the particular national government. Site surveys usually give a more detailed information of the site conditions to allow power calculation to be done and design work to begin. Flow data should however be collected over a period of one year where possible, this is to ascertain on the fluctuation in the river flow over the various seasons.

Micro Hydro Power Plant Development - Barriers

There are various barriers that hinder the dissemination of MHP, some of them have been identified as:

- Policy and regulatory framework: In most cases there exists no sufficient policies and frameworks that govern MHP schemes, this is because the MHP is either not regulated at all or is combined with a broader framework made for rural electrification which may be unclear and in transparent. Such challenges causes the MHP project developers not to know which requirements apply and work in an unreliable grey area of regulation.

- Financing: Lack of sufficient funding to be used in development is a common challenge as most MHP rely on donor funding which in most cases is only available in funding a small portion of the hydro power potential. One of the ways that such as case can be addressed is if there can be an option of exploring other sources of funding especially from private venture capitalists and local banks.

- Capacity to plan, build and operate MHP plants: Lack of knowledge and awareness on MHP potential posses a great challenge for rural electrification, hydro power schemes still dominate as political decision makers still tend to go for them as a more "modern" approach. Combined with that there is minimal capacity to design, implement and revise the MHP supportive policies and regulations. And at the technical level, local capacity is often missing to plan, build and run MHP projects. The is also a problem in the lack of a ready supply of affordable turbine parts and the lack of domestic manufacturing capacity for hydro systems of all sizes also poses a barrier to a swift and cost-effective MHP project development.

- Data on hydro resources: There is usually a lack of interest in MHP deployment from the politicians and power utilities companied by the lack appropriate capacities and budgets, as well as unavailability of pubic data on MHP sites. Such a lack of sound basic data (e.g. on mid-to long-term hydrological, geographic, geologic data and figures on the current and future demand for electricity and social infrastructure, but especially on effects of seasonal and long-term river flow variations), poses a major barrier for private investors in MHP. This causes a bottleneck for investment in hydropower systems as there is an increase in climate variability accompanied with the destruction of rainfall catchment areas.

Application: Use of Micro Hydro Power Plants

Power produced from a small hydro station can be used for various purposes, some of the uses have been classified as follows:

1. Productive Use: This is where the electricity generated is used to perform activities where money is exchanged for a service. Most of this scenarios take place in small businesses.

2. Consumptive Use: All the other used that the electricity can be used for are called consumptive use. They include using the electricity at the household or close to the household.

Besides the productive and consumptive use, a distinction can also be made between the use of power in a mechanical way or in the form of electricity:

	mechanic	electricity
productive use	<ul><li>agro processing</li><li>timber sawing</li><li>textile fabrication</li><li>cooling</li><li>drying</li></ul>	<ul><li>mechanical uses with electricity as intermediate</li><li>heating</li><li>lighting</li><li>fertiliser production</li></ul>
consumptive use		<ul><li>domestic lighting</li><li>cooking</li><li>cooling</li><li>radio and television</li></ul>

As the above illustration shows power that is generated by MPH is a convenient source of electricity to fuel anything from workshop machines to domestic lighting as the power can also be supplied to villages via portable rechargeable batteries and thus there are no expensive connection costs. Batteries can as well be charged and used to provide the local community with power. For industrial use however, the turbine shaft can be used directly as mechanical power as opposed to converting it into electricity via generator or batteries. This is suitable for agro-processing activities such as milling, oil extraction and carpentry.

Micro Hydro Pros – Advantages

Efficient Energy Source

It only takes a small amount of flow (as little as two gallons per minute) or a drop as low as two feet to generate electricity with micro hydro. Electricity can be delivered as far as a mile away to the location where it is being used.

Reliable Electricity Source

Hydro produces a continuous supply of electrical energy in comparison to other small-scale renewable technologies. The peak energy season is during the winter months when large quantities of electricity are required.

No Reservoir Required

Microhydro is considered to function as a 'run-of-river' system, meaning that the water passing through the generator is directed back into the stream with relatively little impact on the surrounding ecology.

Cost Effective Energy Solution

Building a small-scale hydro-power system can cost from $1,000 – $20,000, depending on site electricity requirements and location. Maintenance fees are relatively small in comparison to other technologies.

Power for Developing Countries

Because of the low-cost versatility and longevity of micro hydro, developing countries can manufacture and implement the technology to help supply much needed electricity to small communities and villages.

Integrate with the Local Power Grid

If your site produces a large amount of excess energy, some power companies will buy back your electricity overflow. You also have the ability to supplement your level of micro power with intake from the power grid.

Micro Hydro Cons – Disadvantages

Suitable Site Characteristics Required

In order to take full advantage of the electrical potential of small streams, a suitable site is needed. Factors to consider are: distance from the power source to the location where energy is required, stream size (including flow rate, output and drop), and a balance of system components — inverter, batteries, controller, transmission line and pipelines.

Energy Expansion not Possible

The size and flow of small streams may restrict future site expansion as the power demand increases.

Low-power in the Summer Months

In many locations stream size will fluctuate seasonally. During the summer months there will likely be less flow and therefore less power output. Advanced planning and research will be needed to ensure adequate energy requirements are met.

Environmental Impact

The ecological impact of small-scale hydro is minimal; however the low-level environmental effects must be taken into consideration before construction begins. Stream water will be diverted away from a portion of the stream, and proper caution must be exercised to ensure there will be no damaging impact on the local ecology or civil infrastructure.

Small Hydro

Small scale hydropower systems capture the energy in flowing water and convert it to usable energy. Although the potential for small hydro-electric systems depends on the availability of suitable water flow, where the resource exists it can provide cheap clean reliable electricity. A well designed small hydropower system can blend with its surroundings and have minimal negative environmental impacts.

Moreover, small hydropower has a huge, as yet untapped potential in most areas of the world and can make a significant contribution to future energy needs. It depends largely on already proven and developed technology, yet there is considerable scope for development and optimization of this technology.

Working of Small Hydro

Hydropower systems use the energy in flowing water to produce electricity or mechanical energy. The water flows via channel or penstock to a waterwheel or turbine where it strikes the bucket of the wheel, causing the shaft o f the waterwheel or turbine to rotate. When generating electricity, the rotating shaft, which is connected to an alternator or generator, converts the motion of the shaft into electrical energy. This electrical energy may be used directly, stored in batteries, or inverted to produce utility-quality electricity. A small scale hydroelectric facility requires that a sizable flow of water and a proper height of fall of water, called head, is obtained without building elaborate and expensive facilities. Small hydroelectric plants can be developed at existing dams and have been constructed in connection with river and lake water-level control and irrigation schemes. By using existing structures, only minor new civil engineering works are required, which reduces the cost of this component of a development.

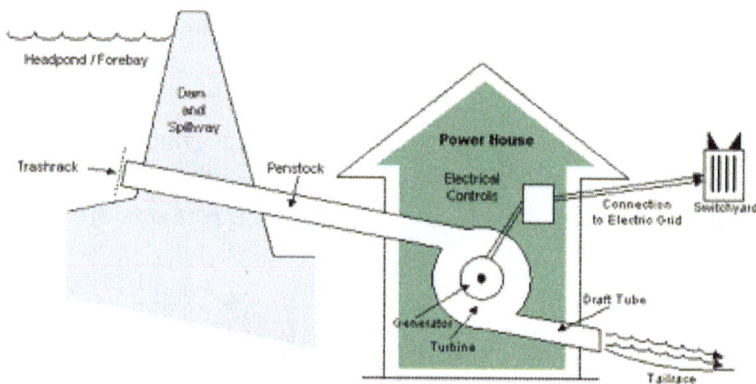

Parts of a Small Scale Hydropower Facility

In other, more rugged regions of the country, it is possible to develop relatively higher heads without elaborate or expensive civil engineering works so that relatively smaller flows are required to develop the desired power. In these cases, it may be possible to construct a relatively simple diversion structure and obtain the highest drop by diverting flows at the top of a waterfall or steeply falling watercourse.

Project Design

Many companies offer standardized turbine generator packages in the approximate size range of 200 kW to 10 MW. These "water to wire" packages simplify the planning and development of the site since one vendor looks after most of the equipment supply. Because non-recurring engineering costs are minimized and development cost is spread over multiple units, the cost of such package systems is reduced. While synchronous generators capable of isolated plant operation are often used, small hydro plants connected to an electrical grid system can use economical induction generators to further reduce installation cost and simplify control and operation.

Small "run of the river" projects do not have a conventional dam with a reservoir, only a weir to form a headpond for diversion of inlet water to the turbine. Unused water simply flows over the weir and the headpond may only be capable of a single day's storage, not enough for dry summers or frozen winters when generation may come to a halt. A preferred scenario is to have the inlet in an existing lake.

Countries like India and China have policies in favor of small hydro, and the regulatory process allows for building dams and reservoirs. In North America and Europe the regulatory process is too long and expensive to consider having a dam and a reservoir for a small project.

Small hydro projects usually have faster environmental and licensing procedures, and since the equipment is usually in serial production, standardized and simplified, and the civil works construction is also reduced, the projects may be developed very rapidly. The physically smaller size of equipment makes it easier to transport to remote areas without good road or rail access.

One measure of decreased environmental impact with lakes and reservoirs depends on the balance between stream flow and power production. Reducing water diversions helps the river's ecosystem, but reduces the hydro system's return on Investment (ROI). The hydro system design must strike a balance to maintain both the health of the stream and the economics.

Benefits of Small-scale Hydropower

Hydroelectric systems provide the following general benefits:

- Hydroelectric energy is a continuously renewable electrical energy source.

- Hydroelectric energy is non-polluting - no heat or noxious gases are released.

- Hydroelectric energy has no fuel cost and with low operating and maintenance costs, it is essentially inflation proof.

- Hydroelectric energy technology is a proven technology that offers reliable and flexible operation.

- Hydroelectric stations have a long life and many existing stations have been in operation for more than half a century and are still operating efficiently.

- Hydropower station efficiencies of over 90% are achieved making it the most efficient of energy conversion technologies.

References

- Pelc, Robin; Fujita, Rod M. (November 2002). "Renewable energy from the ocean". Marine Policy. Elsevier. 26 (6): 471–479. doi:10.1016/S0308-597X(02)00045-3.

- Hydroelectricity: newworldencyclopedia.org, Retrieved 10 April 2018

- Hydropower-and-pumped-storage-22104: altenergymag.com, Retrieved 09 May 2018

- Lamb, H. (1994). Hydrodynamics (6th ed.). Cambridge University Press. ISBN 978-0-521-45868-9. §174, p. 260.

- Conduit-hydropower-recovering-energy-one-control-valve-at-a-time-5519: valvemagazine.com, Retrieved 20 May 2018

- Retiere, C. (January 1994). "Tidal power and the aquatic environment of La Rance". Biological Journal of the Linnean Society. Wiley. 51 (1–2): 25–36. doi:10.1111/j.1095-8312.1994.tb00941.x.

- Hydropower-learning-centre/head-and-flow-detailed-review: renewablesfirst.co.uk, Retrieved 20 July 2018

- Tidalenergy: conserve-energy-future.com, Retrieved 15 March 2018

3

Water Turbines

A rotary machine through which kinetic and potential energy of water is converted into mechanical work is called a water turbine. These can be categorized into two groups, namely, impulse turbine and reaction turbine. This chapter closely examines the diverse aspects of wind turbines and their role in hydropower generation.

Hydraulic Turbine

A hydraulic turbine converts the potential energy of a flowing liquid to rotational energy for further use. In principle, there is no restriction on either the liquid or the use for the energy developed. However, in most cases, these are respectively water and electrical generation. Hence, hydraulic turbines have become synonymous with hydro electric power.

The rate of doing work (power) developed in a hydraulic turbine is:

$$W = \Delta_p \, V \eta$$

Where Δp is the drop in total pressure across the turbine, $\dot{V}$ is the volumetric flow rate and η the efficiency of the turbine. It is common practice to quote Δp in terms of the difference between the upstream and downstream total heads, called the turbine head which equals $\Delta p/g\rho$.

The basis of the design of the turbine hydraulic passages is the velocity diagrams at the entry and exit of the turbine rotating element (called the runner). These lead to the *Euler equation* for theoretical torque and to the theoretical *Euler efficiency* of the turbine. Although elemental velocity triangles are employed for preliminary design of the hydraulic passages, for large turbines, model testing is necessary for verification of performance. Because of the cost and time involved in developmental model testing, more recently, a computerized finite element solution of the inviscid flow equations in the hydraulic passages, cross correlated with general data from model test results, is employed for advanced design. In particular, the *efficiency of the hydraulic turbine* must be optimized and established for contractual purposes. The peak efficiency of properly designed large hydraulic turbines can be as high as 95%, with typically every point of improved efficiency involving considerable monetary benefits in operation.

Testing cannot model the losses due to hydraulic friction. Hence model test efficiencies are converted to full scale values with established Reynolds Number based formulas. Confirmatory site efficiency testing of hydraulic turbines is possible using current meters, ultrasonic, salt velocity tracer, water column inertia (Gibson method) and thermodynamic methods to evaluate flow. However, because of cost and the measuring inaccuracy inherent in site testing, there is a general trend to rely solely on model test results.

Depending on the use, the amount of water and level difference available, the power of hydraulic turbines can be a few kilowatts up to hundreds of Megawatts. However, regardless of size, their performance can be equated through similarity laws; hence the applicability of tests on models to predict the performance of large turbines. From non-dimensional considerations the similarity laws are:

$$\text{Head Coefficient} = \Delta_p / (pu^2)$$

$$\text{Flow Coefficient} = V / (uD^2)$$

$$\text{Power Coefficient} = W / (pu^3 D^2)$$

Where W is the power, ρ is the water density, u is the water velocity, and D is a characteristic diameter of the runner from which all other dimensions of the hydraulic passages follow.

Hydraulic turbines are classified according to specific speed. Specific speed is defined as the rotational speed (revolutions per minute) at which a hydraulic turbine would operate at best efficiency under unit head (one meter) and which is sized to produce unit power (one kilowatt). The equation for specific speed derived from non- dimensional considerations is therefore:

$$\text{Specific speed} = \text{Speed}(W / 1000)^{0.5} / (\Delta_p / \rho g)^{125}$$

The historical development of hydraulic turbines has culminated in two distinct types namely *impulse* (or constant pressure) and *reaction*. Reaction turbines are further divided into radial and axial flow and variable and fixed runner blade. In the impulse turbine, flow is directed through a nozzle to impact on a series of buckets attached to the periphery of the runner. The total transfer of energy is from the change of momentum of the fluid jet; there is no change in hydrostatic pressure once the fluid exits the jet. Impulse turbines are typically used for heads above 100 m and reasonably low flows. They can have up to six jets to better utilize larger flows, as shown in figure below.

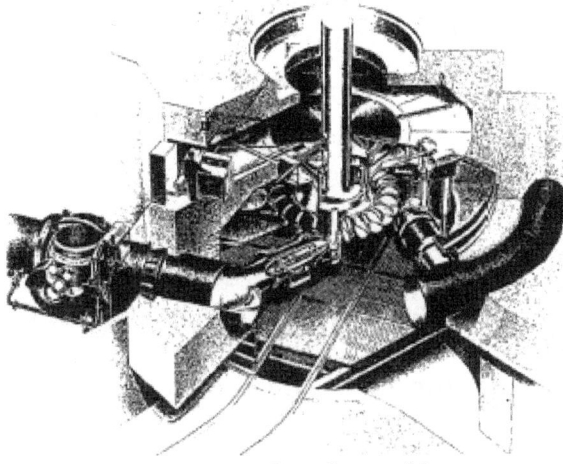

Figure: Four jet Pelton turbine

Reaction turbines for heads in the range 600 m to 30 m are known as *Francis turbines* (after their inventor). These are radial flow units in which the flow enters the runner radially and discharges axially, as illustrated in figure below.

The runner of a reaction turbine is equipped with blades which contain and direct the flow. Therefore, in addition to energy derived from the momentum changes of the fluid as it passes through the runner, it is also generated from the changes in hydrostatic pressure of the fluid within the runner passages. Below design heads of approximately 50 m axial flow turbines are used, known as *propeller units* because of their similarity to a ships propeller. A subsection of propeller units known as *bulb turbines* are used for heads below approximately 10 m. Small hydraulic turbines can be arranged with a horizontal shaft for ease of maintenance, but the larger units used for hydroelectric power installations are almost universally vertical. The exception is the bulb turbine, which is only arranged horizontally.

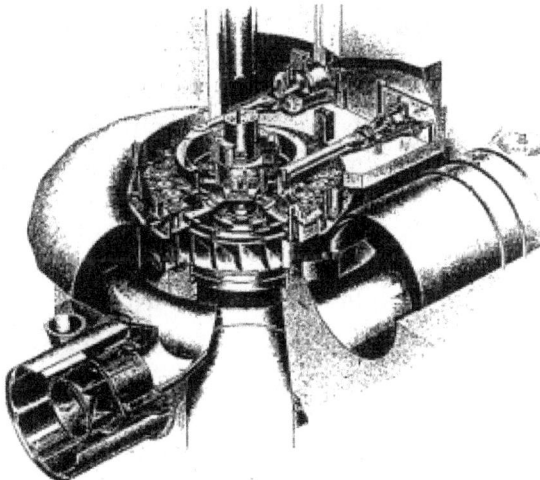

Figure: Francis turbine with pressure relief valve

Utilization of the three basic turbine types are within the following ranges of specific speed (calculated from rpm, kW and m):

- Single jet Pelton: 3 < Specific Speed < 36

- Multiple jet Pelton: 36 < Specific Speed < 60

- Francis: 60 < Specific Speed < 400

- Propeller: 300 < Specific Speed < 1200

Reversible pump turbines are a special type of reaction turbine. These change direction of rotation to operate both as a pump and a turbine and are used for pumped storage applications. Hydraulically, a reversible pump turbine is designed as a pump with only minor modifications to accommodate its role as a turbine.

The stability of operation and the internal hydraulic forces (both static and dynamic), are directly dependent on the velocity of flow through the turbine. For a given design head and flow velocity there is the unique specific speed. This leads to the relationship:

$$\text{Specific speed} = K / \left(\Delta_p / \rho g \right)^{0.5}$$

Where K is a constant.

There are strong commercial benefits in using as small a turbine (hence high flow velocities) as possible for any given application. However, this is restricted by the state of the art in respect of vibration and performance. In 1994, the generally accepted maximum value for K was about 2300.

In all three general groups of hydraulic turbine, flow is directed to the periphery of the runner of the turbine via a spiral casing and discharges from the runner through a draft tube. In reaction turbines, the rotation of the liquid commences as a free vortex in the spiral casing; it is directed through the fixed stay vanes and then through the adjustable turbine wicket gates, such that the angle of approach of the flow to the runner at the design conditions is precisely the runner blade angle. Flow through the reaction turbine to obtain the required power is regulated by the turbine wicket gates. Thus, shockless flow is only obtained at the output for best efficiency and at the design head. In a reaction turbine the draft tube is designed for maximum recovery of hydrostatic pressure. This is especially critical for low head turbines.

The efficiency and operational stability of turbines with low design heads is a strong function of the inlet approach angle, efficiency dropping rapidly with decrease in output and, to a lesser extent, with change in head. To maintain efficiency over the operating range, low head units often have adjustable runner blades, the runner inlet angle changing with wicket gate position and, if required, with operating head. These units are said to be double regulating and include the semi radial flow *Deriaz turbines*, axial flow *Kaplan turbines* (both named after their inventors) and bulb turbines.

Reaction turbine runners can suffer Cavitation at the blade inlet (due to off design flow conditions), in the runner hydraulic channels at part load operation and at the runner exit on the suction side of the runner blades. The latter is the most critical and is a function of the back pressure on the runner. Suction side cavitation in a hydraulic turbine is accordingly related to downstream (tailwater) level through the Thoma Co-efficient defined as:

$$\sigma_{Th} = \left(p_a - p_{vap} - z \cdot gp\right)/\Delta_p$$

Where z is the height of the runner exit plane above tailwater level, p_a is atmospheric pressure and p_{vap} vapor pressure of the fluid. Height z is commonly called the setting of the turbine.

Typically, it is too expensive to set a large hydraulic turbine deep enough below tailwater level to completely eliminate cavitation and, in any particular application, an economic balance between cavitation repair and cost of excavation has to be established. For preliminary design, operating experience is used to establish an acceptable σ_{Th}. One such criterion is:

$$\sigma_{Th} = 7.54.10^{-5} \cdot \left(\text{Specific speed}\right)^{1.41}$$

For major installations, the cavitation performance of the hydraulic turbine should be established with model testing. Cavitation model testing of medium to tow head hydraulic turbines is often conducted with Froude Number similarity to the prototype.

The wetted surfaces of hydraulic turbines are also prone to damage due to **Erosion** by transported silt and sand and corrosion from aggressive fluids. Damage is particularly problematical when silt erosion, corrosion and cavitation act in conjunction (synergistic effects). In applications where the silt content is extreme the hydraulic design may sacrifice efficiency for partial immunity from silt erosion (contouring of surfaces and thickening of runner blades, for example).

Stainless steel, which has a far better resistance to cavitation and erosion than carbon steel, is extensively employed in susceptible areas.

The economic pressures to increase specific speed and hence flow velocities, for a particular head, have led to operational problems with flow induced vibrations from runner blades and stay vanes. These are particularly problematical when their forcing frequency coincides with the natural frequency of any other part of the mechanical, hydraulic or electrical system thus leading to resonance. Also, at part load operation reaction turbines suffer from draft tube pressure oscillations, which can result in unacceptable power swings. Air admitted naturally to areas of low pressure or force fed from compressors can be effective in curing oscillations due to part load operation.

The speed of hydraulic turbines is regulated by a governor. The governor senses the speed of the turbine and adjusts the wicket gate opening to maintain speed within close limits. Speed sensing and the associated feedback control systems are typically digital. On all other than very small hydraulic turbines, the amplification from the digital governor to the wicket gates (and runner blades if required) is through a high pressure oil servomotor system. If the hydraulic turbine is operating on a large integrated network then its speed is controlled by the network and the governor is used to change output via its permanent speed droop. The governor feedback and gain have to accommodate the water column and generator inertias. The turbine and all equipment connected to it, both mechanically and electrically, are designed for runaway speed of the turbine (speed at zero torque) resulting from governor failure. To aid regulation, high to medium head turbines are often equipped with pressure relief valves. For the same reason, the jets of impulse turbines are often equipped with flow diverters. For security, isolation valves or hydraulic gates are commonly installed at spiral casing inlets.

Working Principle of Hydraulic Turbine

According to Newton's law a force is directly proportional to the change in momentum. So if there is any change in momentum of fluid a force is generated. In the hydraulic turbine blades or bucket (in case of Pelton wheel) are provided against the flow of water which change the momentum of it. As the momentum is change a resulting pressure force generated which rotate the rotor or turbine. The most important phenomenon is the amount of change in momentum of water which is directly proportional to force. As the change in momentum high the force generated is high which increase the energy conversion. So the blade or buckets are designed so it can change maximum momentum of water. This is the basic principle of turbine. These turbines are used as hydro electric power plant.

Inside a Hydropower Plant

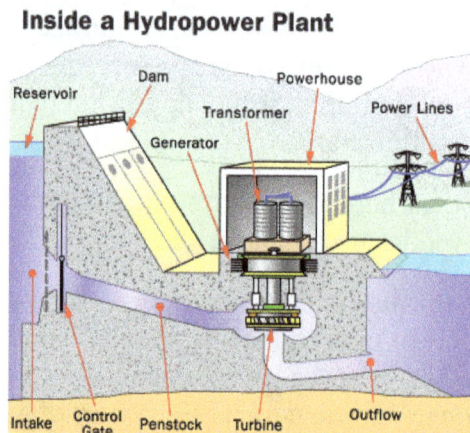

Advantages and Disadvantages:

Hydro power plant or we can say that hydraulic turbines are widely used form the last decades. It is an efficient renewable energy source. There are many up and downs in every project so there are also have many advantages and disadvantages which are describe below:

Advantages

- It is a renewable energy source. Water energy can be used again and again.

- The running cost of turbine is less compare to other.

- It has high efficiency.

- It can be control fully. The gate of dam is closed when we does not need electricity and can be open when we needed.

- Dams are used from very long time so it can be used for power generation.

- It does not pollute environment.

- It is easy to maintain.

- The dam constructed for hydraulic turbine can become a tourist place.

Disadvantages

- Initial cost is very high. It takes several decades to become profitable.

- It can destroy the natural environment at site. Large dam cause big geological damages.

- It can develop at only few sites where proper amount of water is available.

Water Wheel

Waterwheel is a mechanical device for tapping the energy of running or falling water by means of a set of paddles mounted around a wheel. The force of the moving water is exerted against the paddles, and the consequent rotation of the wheel is transmitted to machinery via the shaft of the wheel. The waterwheel was perhaps the earliest source of mechanical energy to replace that of humans and animals, and it was first exploited for such tasks as raising water, fulling cloth, and grinding grain.

The combination of waterwheel and transmission linkage, often including gearing, was from the Middle Ages usually designated a mill. Of the three distinct types of water mills, the simplest and probably the earliest was a vertical wheel with paddles on which the force of the stream acted. Next was the horizontal wheel used for driving a millstone through a vertical shaft attached directly to the wheel. Third was the geared mill driven by a vertical waterwheel with a horizontal shaft. This required more knowledge and engineering skill than the first two, but it had much greater potential. Vertical waterwheels were also distinguished by the location of water contact with the wheel: first, the undershot wheel; second, the breast wheel; and third, the overshot wheel. These waterwheels generally used the energy of moving streams, but tidal mills also appeared in the 11th century.

Each type of mill had its particular advantages and disadvantages. Relatively little is known of their development before the Middle Ages, but certain of their characteristics suggest an order of appearance within the context of the complexity of construction and the possibilities for utilization.

The simple vertical wheel required little extra structure, but the force and rate of power takeoff were dependent upon stream characteristics and wheel diameter. Since change of power direction was not involved, this wheel proved most useful in raising water, utilizing, for instance, a string of pots worked by a chain drive.

The horizontal-wheel mill (sometimes called a Norse or Greek mill) also required little auxiliaryconstruction, but it was suited for grinding because the upper millstone was fixed upon the vertical shaft. The mill, however, could only be used where the current flow was suitable for grinding.

The geared vertical-wheel mill was more versatile. Construction was relatively simple if the wheel was of the undershot kind, because the wheel paddles could be simply dipped in the stream flow, whether it was river, tide, or man-built millrace. A millwright could choose his gear ratio to match power utilization with rate of stream flow, and the wheel could be mounted in a bridge arch or on a barge anchored in midstream. Vitruvius described the first geared vertical wheel for which we have good evidence. This mill is also of major significance because it was the first application of gearing to utilize other than muscle power. This mill had an undershot wheel and, unlike the breast or overshot wheels, did not make use of the weight of falling water.

Ḥamāh Wooden waterwheels

Mills with geared breast and overshot wheels required more auxiliary construction, but they allowed the most generalized exploitation of available water power. A major construction problem was locating a mill where the fall of water would be suited to the desired diameter of the wheel. Either a long millrace from upstream or a dam could be used.

Little is known of the details of geared-mill development between the time of Vitruvius and the 12th century. An outstanding installation was the grain mill at Barbegal, near Arles, France, which had 16 cascaded overshot wheels, each 7 feet (2 metres) in diameter, with wooden gearing. It is estimated that this mill could meet the needs of a population of 80,000.

Even though the highly adaptable, geared mill, with its widely diversified stream-flow conditions, was used in the Roman Empire, historical evidence suggests that its most dramatic industrial consequences occurred during the Middle Ages in Western Europe. After the 13th century the overshot waterwheel appears to have become more common than the undershot wheel.

The geared mill of the Middle Ages was actually a general mechanism for the utilization of power. The power from a horse- or cattle-powered mill was small compared to that from overshot water-wheels, which usually generated two to five horsepower.

Stream

Stream wheel

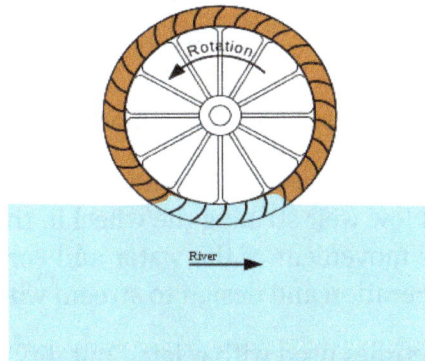

Diagram of stream shot waterwheel.

A stream wheel is a vertically mounted water wheel that is rotated by the water in a water course striking paddles or blades at the bottom of the wheel. This type of water wheel is the oldest type of horizontal axis wheel. They are also known as free surface wheels because the water is not constrained by millraces or wheel pit.

Stream wheels are cheaper and simpler to build, and have less of an environmental impact, than other type of wheel. They do not constitute a major change of the river. Their disadvantages are their low efficiency, which means that they generate less power and can only be used where the flow rate is sufficient. A typical flat board undershot wheel uses about 20 percent of the energy in the flow of water striking the wheel as measured by English civil engineer John Smeaton in the 18th century. More modern wheels have higher efficiencies.

Stream wheels gain little or no advantage from head, a difference in water level.

Stream wheels mounted on floating platforms are often referred to as ship wheels and the mill as a ship mill. The earliest were probably constructed by the Byzantine general Belisarius during the siege of Rome in 537. Later they were sometimes mounted immediately downstream from bridges where the flow restriction of the bridge piers increased the speed of the current.

Historically they were very inefficient but major advances were made in the eighteenth century.

Undershot Wheel

Undershot waterwheel

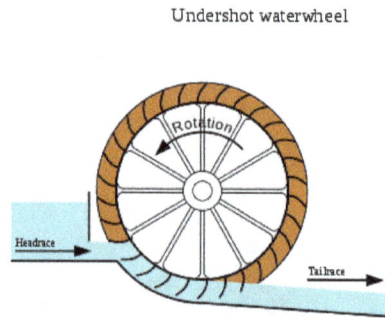

Diagram of undershot waterwheel showing headrace, tailrace and water.

An undershot wheel is a vertically mounted water wheel with a horizontal axle that is rotated by the water from a low weir striking the wheel in the bottom quarter. Most of the energy gain is from the movement of the water and comparatively little from the head. They are similar in operation and design to stream wheels.

The term undershot is sometimes used with related but different meanings:

- All wheels where the water passes under the wheel.
- Wheels where the water enters in the bottom quarter.
- Wheels where paddles are placed into the flow of a stream. See stream above.

This is the oldest type of vertical water wheel.

Breastshot Wheel

The word breastshot is used in a variety of ways. Some authors restrict the term to wheels where the water enters at about the 10 o'clock position, others 9 o'clock, and others for a range of heights. In this article it is used for wheels where the water entry is significantly above the bottom and significantly below the top, typically the middle half.

Breastshot waterwheel

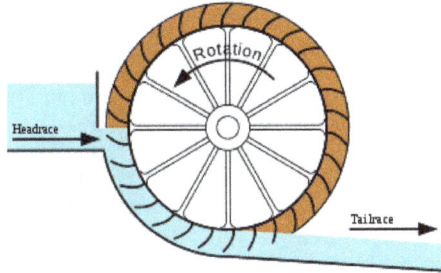

Diagram of breastshot waterwheel showing headrace, tailrace and water.

They are characterised by:

- Buckets carefully shaped to minimise turbulence as water enters.

- Buckets ventilated with holes in the side to allow air to escape as the water enters.

- A masonry "apron" closely conforming to the wheel face, which helps contain the water in the buckets as they progress downwards.

Both kinetic (movement) and potential (height and weight) energy are utilised.

The small clearance between the wheel and the masonry requires that a breastshot wheel has a good trash rack ('screen' in British English) to prevent debris from jamming between the wheel and the apron and potentially causing serious damage.

Breastshot wheels are less efficient than overshot and backshot wheels but they can handle high flow rates and consequently high power. They are preferred for steady, high-volume flows such as are found on the Fall Line of the North American East Coast. Breastshot wheels are the most common type in the United States of America and are said to have powered the industrial revolution.

Backshot Wheel

A backshot wheel (also called pitchback) is a variety of overshot wheel where the water is introduced just before the summit of the wheel. In many situations it has the advantage that the bottom of the wheel is moving in the same direction as the water in the tail race which makes it more efficient. It also performs better than an overshot wheel in flood conditions when the water level may submerge the bottom of the wheel. It will continue to rotate until the water in the wheel pit rises quite high on the wheel. This makes the technique particularly suitable for streams that experience significant variations in flow and reduces the size, complexity and hence cost of the tail race.

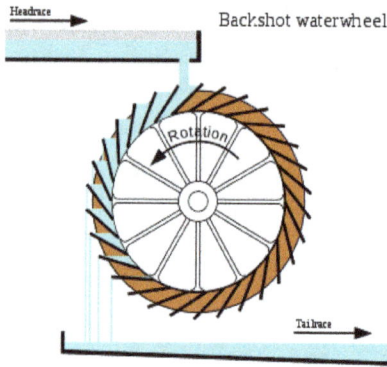

Diagram of backshot waterwheel showing headrace, tailrace, water, and spillage.

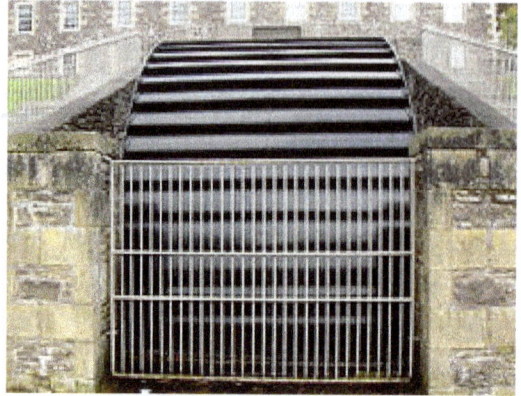

Backshot wheel

The direction of rotation of a backshot wheel is the same as that of a breastshot wheel but in other respects it is very similar to the overshot wheel.

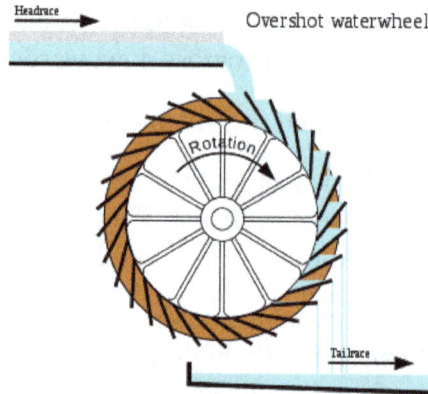

Overshot Wheel

Diagram of overshot waterwheel showing headrace, tailrace, water, and spillage.

A vertically mounted water wheel that is rotated by water entering buckets just past the top of the wheel is said to be overshot. The term is sometimes, erroneously, applied to backshot wheels where the water goes down behind the wheel.

A typical overshot wheel has the water channelled to the wheel at the top and slightly beyond the axle. The water collects in the buckets on that side of the wheel, making it heavier than the other "empty" side. The weight turns the wheel, and the water flows out into the tail-water when the wheel rotates enough to invert the buckets. The overshot design is very efficient, it can achieve 90%, and does not require rapid flow.

Nearly all of the energy is gained from the weight of water lowered to the tail race although a small contribution may be made by the kinetic energy of the water entering the wheel. They are suited to larger heads than the other type of wheel so they are ideally suited to hilly country. However even the largest water wheel, the Laxey Wheel in the Isle of Man, only utilises a head of ~30m. The world's largest head turbines, Bieudron Hydroelectric Power Station in Switzerland, utilise ~1869m.

Overshot wheels require a large head compared to other types of wheel which usually means significant investment in constructing the head race. Sometimes the final approach of the water to the wheel is along a flume or penstock, which can be lengthy.

Hybrid

Overshot and Backshot

One of Finch Foundry's water wheels

Some wheels are overshot at the top and backshot at the bottom thereby potentially combining the best features of both types. The photograph shows an example at Finch Foundry in Devon, UK. The head race is the overhead timber structure and a branch to the left supplies water to the wheel. The water exits from under the wheel back into the stream.

Reversible

The Anderson Mill of Texas is undershot, backshot, and overshot using two sources of water.
This allows the direction of the wheel to be reversed.

A special type of overshot/backshot wheel is the reversible water wheel. This has two sets of blades or buckets running in opposite directions, so that it can turn in either direction depending on which side the water is directed. Reversible wheels were used in the mining industry in order to power various means of ore conveyance. By changing the direction of the wheel, barrels or baskets of ore could be lifted up or lowered down a shaft or inclined plane. There was usually a cable drum or a chain basket (German: Kettenkorb) on the axle of the wheel. It is essential that the wheel have braking equipment to be able to stop the wheel (known as a braking wheel). The oldest known drawing of a reversible water wheel was by Georgius Agricola and dates to 1556.

Suspension Wheels and Rim-gears

The suspension wheel with rim-gearing at the Portland Basin Canal Warehouse

Two early improvements were suspension wheels and rim gearing. Suspension wheels are constructed in the same manner as a bicycle wheel, the rim being supported under tension from the hub- this led to larger lighter wheels than the former design where the heavy spokes were under compression. Rim-gearing entailed adding a notched wheel to the rim or shroud of the wheel. A stub gear engaged the rim-gear and took the power into the mill using an independent line shaft. This removed the rotative stress from the axle which could thus be lighter, and also allowed more flexibility in the location of the power train. The shaft rotation was geared up from that of the wheel which led to less power loss. An example of this design pioneered by Thomas Hewes and refined by William Fairburn can be seen at the 1849 restored wheel at the Portland Basin Canal Warehouse.

Efficiency

Overshot (and particularly backshot) wheels are the most efficient type; a backshot steel wheel can be more efficient (about 60%) than all but the most advanced and well-constructed turbines. In some situations an overshot wheel is preferable to a turbine.

The development of the hydraulic turbine wheels with their improved efficiency (>67%) opened up an alternative path for the installation of water wheels in existing mills, or redevelopment of abandoned mills.

The Power of a Wheel

The energy available to the wheel has two components:

- Kinetic energy: Depends on how fast the water is moving when it enters the wheel.

- Potential energy: Depends on the change in height of the water between entry to and exit from the wheel.

The kinetic energy can be accounted for by converting it into an equivalent head, the velocity head, and adding it to the actual head. For still water the velocity head is zero, and to a good approximation it is negligible for slowly moving water, and can be ignored. The velocity in the tail race is not taken into account because for a perfect wheel the water would leave with zero energy which requires zero velocity. That is impossible, the water has to move away from the wheel, and represents an unavoidable cause of inefficiency.

The power is how fast that energy is delivered which is determined by the flow rate.

Quantities and Units

- η = efficiency

- ρ = density of water (1000 kg/m³)

- A = cross sectional area of the channel (m²)

- D = diameter of wheel (m)

- P = power (W)

- d = distance (m)

- g = strength of gravity (9.81 m/s² = 9.81 N/kg)

- h = head (m)

- h_p = pressure head, the difference in water levels (m)

- h_v = velocity head (m)

- k = velocity correction factor. 0.9 for smooth channels.

- v = velocity (m/s)

- $\dot{q}$ = volume flow rate (m³/s)

- t = time (s)

Measurements

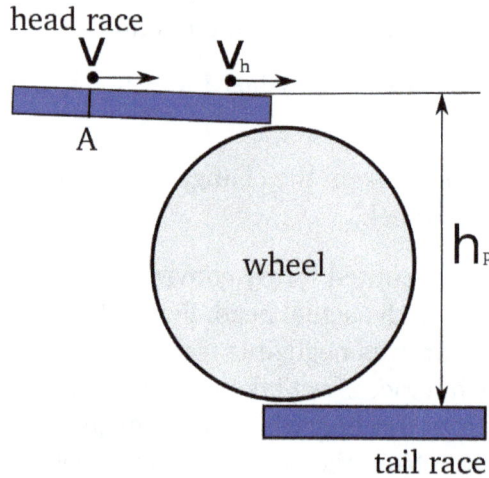

How to measure the head and flow rate of a water wheel

Pressure head h_p = is the difference in height between the head race and tail race water surfaces.

Velocity head h_v = is calculated from the velocity of the water in the head race at the same place as the pressure head is measured from.

The velocity (speed) v can be measured by the pooh sticks method, timing a floating object over a measured distance. The water at the surface moves faster than water nearer to the bottom and sides so a correction factor should be applied as in the formula below.

There are many ways to measure the volume flow rate. Two of the simplest are:

- From the cross sectional area and the velocity. They must be measured at the same place but that can be anywhere in the head or tail races. It must have the same amount of water going through it as the wheel.

- It is sometimes practicable to measure the volume flow rate by the bucket and stop watch method.

Formulae

Quantity	Formula
Power	$P = \eta \cdot \rho \cdot g \cdot h \cdot \dot{q}$

Effective head	$h = h_p + h_v$
Velocity head	$\dfrac{\rho \cdot v^2}{2 \cdot a}$
Volume flow rate	$\dot{q} = A \cdot v$
Water velocity (speed)	$v = k \cdot \dfrac{d}{t}$

Rules of Thumb

Breast and Overshot

Quantity	Approximate Formula
Power (assuming 70% efficiency)	$P = 7000 \cdot \dot{q} \cdot h$
Optimal rotational speed	$\dfrac{21}{\sqrt{D}}$ rpm

Traditional Undershot Wheels

Quantity	Approximate Formula
Power (assuming 20% efficiency)	$P = 100 \cdot A \cdot v$
Optimal rotational speed	$\dfrac{9 \cdot v}{D}$ rpm

Hydraulic Wheel Part Reaction Turbine

A parallel development is the hydraulic wheel/part reaction turbine that also incorporates a weir into the centre of the wheel but uses blades angled to the water flow. The WICON-Stem Pressure Machine (SPM) exploits this flow. Estimated efficiency 67%.

The University of Southampton School of Civil Engineering and the Environment in the UK has investigated both types of Hydraulic wheel machines and has estimated their hydraulic efficiency and suggested improvements, i.e. The Rotary Hydraulic Pressure Machine. (Estimated maximum efficiency 85%).

These type of water wheels have high efficiency at part loads / variable flows and can operate at very low heads, < 1 metre. Combined with direct drive Axial Flux Permanent Magnet Alternators and power electronics they offer a viable alternative for low head hydroelectric power generation.

Water-lifting

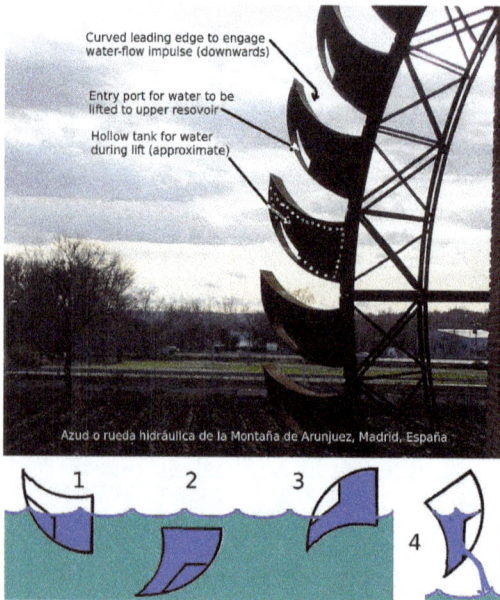

Curved leading edge to engage
water-flow impulse (downwards)

Entry port for water to be
lifted to upper resovoir

Hollow tank for water
during lift (approximate)

Azud o rueda hidráulica de la Montaña de Arunjuez, Madrid, España

ELEVATION

SUMP

PLAN

Detail of *azud* at Aranjuez, Spain   Drainage wheel from Rio Tinto mines

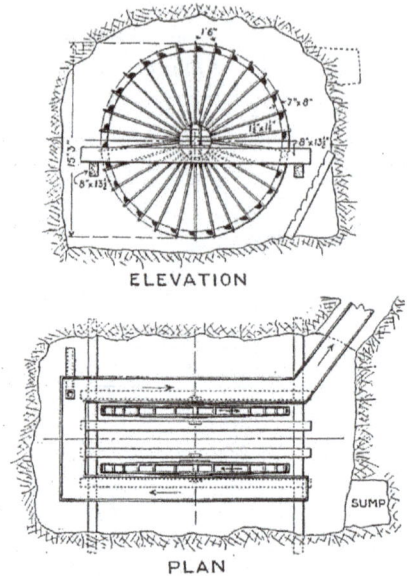

In water-raising devices rotary motion is typically more efficient than machines based on oscillating motion.

The compartmented water wheel comes in two basic forms, the wheel with compartmented body (Latin *tympanum*) and the wheel with compartmented rim or a rim with separate, attached containers. The wheels could be either turned by the flow of water, men treading on its outside or by animals by means of a sakia gear. While the tympanum had a large discharge capacity, it could lift the water only to less than the height of its own radius and required a large torque for rotating. These constructional deficiencies were overcome by the wheel with a compartmented rim which was a less heavy design with a higher lift.

Ptolemaic Egypt

The earliest literary reference to a water-driven, compartmented wheel appears in the technical treatise *Pneumatica* of the Greek engineer Philo of Byzantium. In his *Parasceuastica*, Philo advises the use of such wheels for submerging siege mines as a defensive measure against enemy sapping. Compartmented wheels appear to have been the means of choice for draining dry docks in Alexandria under the reign of Ptolemy IV. The non-existence of the device in the Ancient Near East before Alexander's conquest can be deduced from its pronounced absence from the otherwise rich oriental iconography on irrigation practices. Unlike other water-lifting devices and pumps of the period though, the invention of the compartmented wheel cannot be traced to any particular Hellenistic engineer and may have been made in the late 4th century BC in a rural context away from the metropolis of Alexandria.

The earliest depiction of a compartmented wheel is from a tomb painting in Ptolemaic Egypt which dates to the 2nd century BC. It shows a pair of yoked oxen driving the wheel via a sakia gear, which is here for the first time attested, too. The Greek sakia gear system is already shown fully developed to the point that "modern Egyptian devices are virtually identical". It is assumed that the scientists of the Museum of Alexandria, at the time the most active Greek research center, may have been involved in its invention. An episode from the Alexandrian War in 48 BC tells of how Caesar's enemies employed geared water wheels to pour sea water from elevated places on the position of the trapped Romans.

Around 300 AD, the noria was finally introduced when the wooden compartments were replaced with inexpensive ceramic pots that were tied to the outside of an open-framed wheel.

Impulse Turbine

An impulse turbine is a turbine that is driven by high velocity jets of water or steam from a nozzle directed on to vanes or buckets attached to a wheel. The resulting impulse (as described by Newton's second law of motion) spins the turbine and removes kinetic energy from the fluid flow. Before reaching the turbine the fluid's pressure head is changed to velocity head by accelerating the fluid through a nozzle. This preparation of the fluid jet means that no pressure casement is needed around an impulse turbine.

Working Principle

The potential energy of the water is converted into kinetic energy by passing it through a nozzle. Once we have high speed water jet, we can use its impact to rotate a turbine.

Or we can say, it works on Newton's second law of motion, that it depends on two main factors, mass of water flowing in-to turbine, and change in the velocity of the flow coming in-to turbine to that of going out of turbine after impact. As the mass of water entering into the turbine is same as the water going out of turbine after impact, but with a considerable decrement in its velocity. And the intensity of impact depends upon the time taken by velocity to change from maximum (jet velocity) to minimum. Thus impulse turbine only uses the kinetic energy of water to get its power.

Components of Impulse Turbine

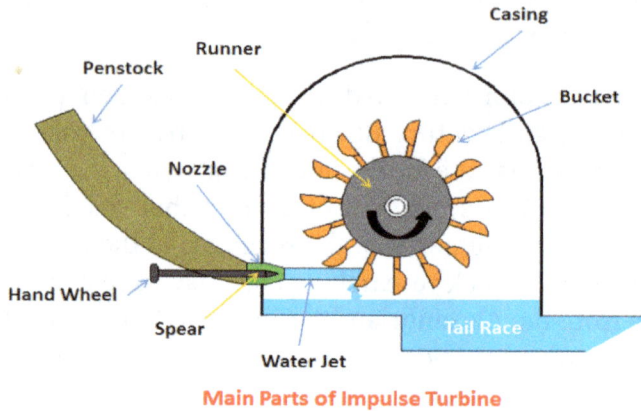

Main Parts of Impulse Turbine

Runner

It is a solid circular-disc with cylindrical shaft in the center. The shaft and the runner both are made from high strength stainless steel where load on the turbine is considerably high. Runners are also made from cast iron where available water head is a bit low, thus force on turbine is not that high.

Buckets

Buckets are cup type hollow hemispherical structures, bolted on the periphery of the runner. Jet strikes these buckets to rotate the runner. Their design plays a vital role in deciding the efficiency of a turbine. These are made either from stainless-steel or cast iron.

Nozzle and Spear

Nozzle directs the flow of water to the buckets, with an increased velocity coming from a high head. Spear is a conical structure which is moved in and out of nozzle to regulate the flow of water striking the buckets.

Casing

Casing of an impulse turbine is a preventive shielding over the turbine, usually made of cast iron. It also prevents the water from splashing, and also guides it to the spill way.

Working of Impulse Turbine

Water stored at a height is passed through a nozzle, situated almost at ground level or ever below ground level. Thus converting the energy of stored water into high speed jet. This high speed water jet strikes the buckets or blades attached to the runner, forcing runner to rotate at its own axis. Thus converting the energy of high speed-jet into rotational energy. This rotational movement of turbine shaft is used to produce electricity through generator.

Spear is moved in and out of nozzle to regulate the flow of water, according to the load on turbine. To get maximum power output from a turbine the velocity of jet striking the buckets should be as much as twice the speed of rotating buckets. So velocity of water-jet is regulated according to the load or rpm of turbine in such a way that we can keep turbine running in its most efficient range.

Moreover, practically we use 3 to 4 nozzles instead of one. This is done to deal with the high loads on turbine and to increase the power output capability of a turbine. Power is also regulated by closing few nozzles when load on the turbine is low.

When the load on the turbine decreases suddenly, and spear could not act fast enough to regulate the flow of water-jet, the rpm of turbine will keep on increasing and could damage the turbine. To prevent this from happening we use deflector which deflects the flow of water jet away from the turbine buckets. Thus keeping turbine under safe limits.

Applications

- It is used worldwide to produce electrical energy in a number of hydro-power plants.

- Turbochargers in automobiles uses the pressure energy of exhaust gases through impulse turbine. Where hot and pressurized gases coming out of exhaust are converted into high velocity jet by passing them through nozzle.

- It is also used in reverse osmosis plant, where waste water jet velocity is used to run turbine, thus acts as an energy recovery system.

Pelton Turbine

A Pelton turbine or Pelton wheel is a type of turbine used frequently in hydroelectric plants. These turbines are generally used for sites with heads greater than 300 meters. This type of turbine was created during the gold rush in 1880 by Lester Pelton.

When used for generating electricity, there is usually a water reservoir located at some height above the Pelton turbine. The water then flows through the penstock to specialized nozzles that introduce pressurized water to the turbine. To prevent irregularities in pressure, the penstock is fitted with a surge tank that absorbs sudden fluctuations in water that could alter the pressure. Unlike other types of turbines which are reaction turbines, the Pelton turbine is known as an impulse turbine. This simply means that instead of moving as a result of a reaction force, water creates some impulse on the turbine to get it to move.

Design

The Pelton turbine has a fairly simplistic design. A large circular disk is mounted on some sort of rotating shaft known as a rotor. Mounted on this circular disk are cup shaped blades known as buckets evenly spaced around the entire wheel. Generally, the buckets are arranged in pairs around the rim. Then nozzles are arranged the wheel and serve the purpose of introducing water to the turbine. Jets of water emerge from these nozzles, tangential to the wheel of the turbine. This causes the turbine to spin as a result of the impact of the water jets on the buckets.

Operation

The operation of a Pelton turbine is fairly simple. In this type of turbine, high speed jets of water emerge from the nozzles that surround the turbine. These nozzles are arranged so the water jet will hit the buckets at splitters, the center of the bucket where the water jet is divided into two streams. The two separate streams then flow along the inner curve of the bucket and leave in the opposite direction that it came in. This change in momentum of the water creates an impulse on the blades of the turbine, generating torque and rotation in the turbine.

The high speed water jets are created by pushing high pressure water (such as water falling from high heads) through nozzles at atmospheric pressure. The maximum output is obtained from a Pelton turbine when the impulse obtained by the blades is maximum, meaning that the water stream is deflected exactly opposite to the direction at which it strikes the buckets at. As well, the efficiency of these wheels is highest when the speed of the movement of the cups is half of the speed of the water jet.

Jonval Turbine

A Jonval turbine, built in 1885. It was in service for about 100 years in a Geneva pump station, where energy in the form of pressurized water was produced for the local industry. Over pressure in the network was released through the world-famous Jet d'Eau. In total, 17 such turbines were operating in the pump station.

The Jonval turbine is a water turbine design invented in France in 1843, in which water descends through fixed curved guide vanes which direct the flow sideways onto curved vanes on the runner. It is named after Feu Jonval, who invented it. The Jonval incorporated ideas from European mathematicians and engineers, including the use of curved blades. This new turbine failed to satisfy the public interest in seeing the water wheels in action, which was likely accepted as a minor drawback at that time.

This type is efficient at full gate, but at partial gate it is less efficient than a Francis turbine. The usual orientation of the wheel was horizontal and the first devices were even alternatively named as "horizontal water wheels". However, some sources mention turbines with both vertical and horizontal shafts.

N.F. Burnham, an American turbine manufacturer, patented numerous improved designs in the second half of 19th century. His turbines had greater efficiency than the Jonval, especially at partial gate, and fewer maintenance problems.

Turgo Turbine

Turgo Turbine was developed Gilkes Energy Company in 1919. The design of this impulse water turbine is inspired from Pelton turbine. The working principle of this turbine is same as other impulse water turbines. It converts the kinetic energy from the water jet into mechanical energy. The main difference between the Turgo and Pelton turbine is the design of runner buckets. The runner of Turgo turbine is the half slice of Pelton turbine. The runner buckets of Pelton turbine has two curved structures as shown in the figure 1 but the Turgo bucket is the half slice of the Pelton bucket as shown in the figure below.

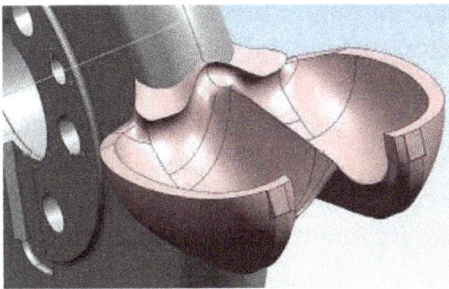

Figure: Bucket of the runner of Pelton turbine   Figure : Buckets of the runner of Turgo turbine

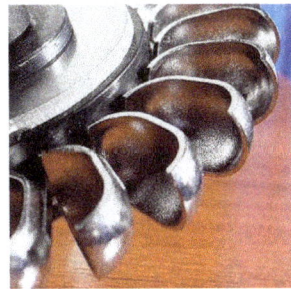

Turgo turbine can handle faster water jet effectively due to its unique bucket design. Buckets in the Pelton turbine don't remove water quickly therefore, water in the bucket interferes with the incoming jet and also reduces the efficiency of the turbine. On the other hand buckets in Turgo turbine remove water from runner quickly. The angel of water jet also plays an important role in this regard. Usually, water jet hits the runner at the angle of 20 degrees. Water jet at this angle provides the maximum impulse to the runner and let the water exit from the other side of the bucket without interfering with incoming water. The working scheme of the Turgo turbine is shown in figure below.

Figure: The working scheme of Turgo turbine

The material of bucket, the diameter of the runner, the width of the nozzle, and height of the head decide the overall effectiveness and efficiency of this turbine. Usually, speed increasing transmission is used with turbines to operate generator at higher speed but Turgo turbine can be connected directly to the generator due to its faster runner. In lab tests, the efficiency of Turgo turbine can be up to 90% but in hydropower plants its efficiency is about 87%.Though the efficiency of Turgo turbine is usually less than that of Pelton turbine but it provides a cost-effective way to generate electricity from medium and small heads and works in the heads of 15 meters to 300 meters. The main reason for less efficiency of Turgo turbine is its less sturdy runner vanes/buckets. Turgo turbine is usually installed in the head where Pelton turbine and Francis turbine overlap. Since Turgo runner is the half slice of Pelton turbine, therefore, Turgo turbine produces the same power as that of Pelton turbine if it has twice the diameter as that Pelton runner. Large diameter leads to higher rotational speed. In some cases, Turgo turbine can even provide better efficiency than Pelton turbine due to its ability to handle the faster flow of water. Turgo turbine can be installed at any site where the Pelton turbine can be used but Pelton turbine can't be used in all applicable sites of Turgo turbines. All these applications of Turgo turbine make it highly desirable for generating electricity.

Reverse overshot Water-wheel

Frequently used in mines and probably elsewhere (such as agricultural drainage), the reverse overshot water wheel was a Roman innovation to help remove water from the lowest levels of underground workings. The remains of such systems found in Roman mines by later mining operations show that they were used in sequences so as to lift water a considerable height.

Vitruvius

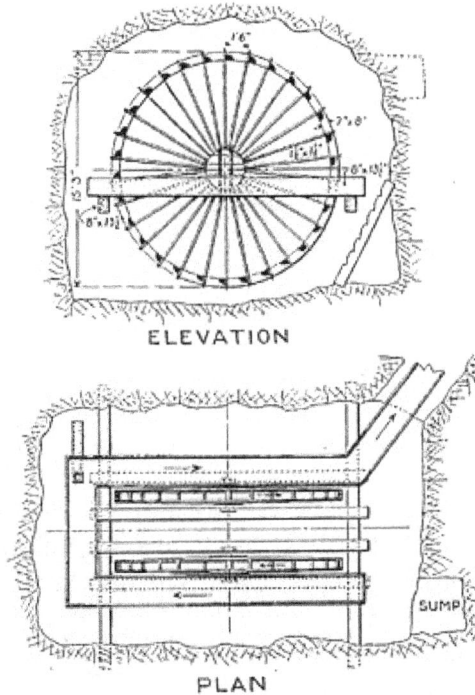

ELEVATION

PLAN

Drainage wheel from Rio Tinto mines

The Roman author Vitruvius gives explicit instructions on the construction of dewatering devices, and describes three variants of the "tympanum" It is a large wheel fitted with boxes, which in the first design, encompass the whole diameter of the wheel. Holes are bored in the boxes to allow water into them, so that as a box dips into the water, it enters and is raised as the wheel turns. When it reaches to the top of the turn, the water runs out into a channel. He then describes a second variant where the boxes are only fitted to the ends of the wheel, so that although the volume of water carried is much smaller, it is carried to a greater height. The final variant is an endless chains of buckets, and much greater lifts can be achieved, although greater effort is needed.

Pliny the Elder

Pliny the Elder is probably referring to such devices in a discussion of silver/lead mines in his Naturalis Historia. Spain produced the most silver in his time, many of the silver mines having been started by Hannibal. One of the largest had galleries running for between one and two miles into the mountain, "water-men" draining the mine, and they *stood night and day in shifts measured by lamps, bailing out water and making a stream.*

That they *stood* suggests that they operated the wheels by standing on the top to turn the cleats, and continuous working would produce a steady stream of water.

Examples

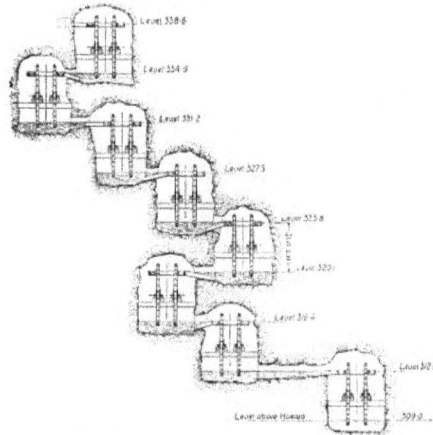

Sequence of wheels found in Rio Tinto mines

Fragments of such machines have been found in mines which were re-opened in the Victorian era in Spain, especially at Rio Tinto, where one example used no less than 16 such wheels working in pairs, each pair of wheels lifting water about 3.5 m (12 feet), so giving a total lift of 30 m (96 feet). The system was carefully engineered, and was worked by individuals treading slats at the side of each wheel. It is not an isolated example, because Oliver Davies mentions examples from the Tharsis copper mine and Logroño in Spain, as well as from Dacia. The gold deposits in Dacia, now modern Romania were especially rich, and worked intensively after the successful Roman invasion under Trajan. According to Oliver Davies, one such sequence discovered at Ruda in Hunedoara County in modern Romania was 75 metres in depth, or over 200 feet. If worked like the Rio Tinto example, it would have needed at least 32 wheels.

One such wheel from Spain was rescued and part of it is now on display in the British Museum. Some of the components are numbered, suggesting that it was prefabricated above ground before assembly in the underground passages. In the 1930s, a fragment of a wooden bucket from a drainage wheel was found in deep workings at the Dolaucothi gold mine in west Wales, and is now preserved in the National Museum of Wales in Cardiff. It has been carbon dated to about 90 AD. From the depth of 160 feet below known open workings, it can be inferred that the drainage wheel was part of a sequence just like that found in Spain. The shape of the edge of one of the lifting buckets is almost identical with that from Spain, suggesting that a template was used to make the devices.

The Cochlea

Another device which was used widely was the Archimedean screw, and examples of such drainage machines have also been found in many old mines. Depictions show the screws being powered by a human treading on the outer casing to turn the entire apparatus as one piece. They were also used in series, so increasing the lift of water from the workings. However, they must have been more difficult to operate since the user had to stand on a slanting surface to turn the screw. The steeper the incline, the greater the risk of the user slipping from the top of the screw. No doubt the reverse water wheel was easier to use with a horizontal treading surface. On the other hand, the screw could be operated by a crank handle fitted to the central axle, but would be more tiring since the weight of the operator does not bear on the crank, as it does when trod from above.

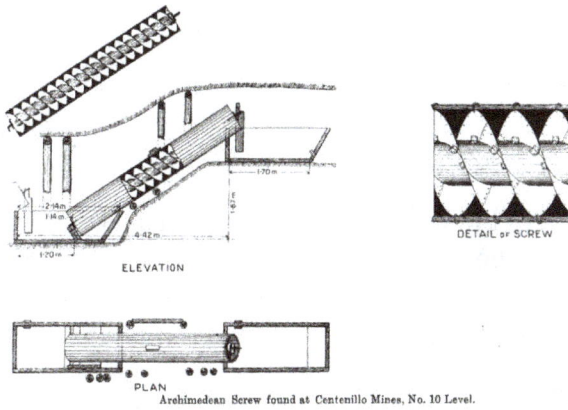

Roman screw used to dewater mines in Spain

Like the reverse water wheel, the cochlea was used for many other purposes apart from draining mines. Irrigation of farmland would have been the most popular application, but any activity which involved lifting water would have employed the devices.

Water Wheels

Multiple sequences of water wheels were used elsewhere in the Roman Empire, such as the famous example at Barbegal in southern France. This system was also a stack of 16 wheels but worked like a normal overshot wheel, the wheels driving stone mills and used to grind grains. The water mills were worked from a masonry aqueduct supplying the Roman town at Arles, and the remains of the masonry mills are still visible on the ground today, unlike the underground drainage systems of the mines, which were destroyed by later mining operations. Other such sequences of mills existed on the Janiculum in Rome, but have been covered and changed by later buildings built on top of them.

Archimedean Screw Hydro Turbine

The Archimedean screw hydro turbine is a relative newcomer to the small-scale hydro world having only arrived on the scene over the last ten years. However, they have been around for many decades as pumps where tens-of-thousands have been installed worldwide, particularly in sewage treatment works. The same manufacturers that dominate the pump market are now the main suppliers into the hydropower market as well.

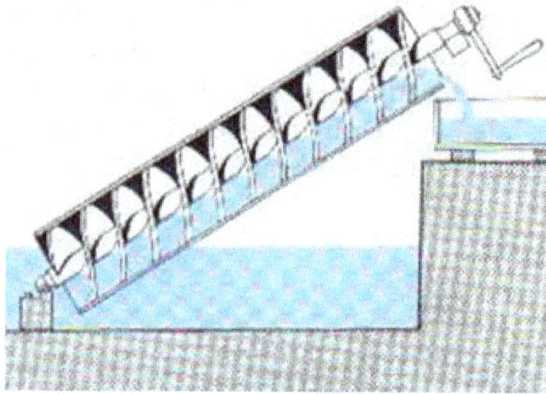

Basic layout of an Archimedean screw pump

As the name suggests, Archimedes is widely acknowledged as the inventor of the screw back in 250 BC, though the credit has been wrongly attributed because they were actually in use in Egypt many years before then. Historically they were used in irrigation to lift water to a higher level and were generally powered by oxen, or even humans on smaller versions. The basic principle of an Archimedean screw pump is shown in the diagram above. If the handle at the top was turned in an anti-clockwise direction it would draw the water up from the lower level to the top.

When used as a hydro turbine the principle is the same but acts in reverse. The water enters the screw at the top and the weight of the water pushes on the helical flights, allowing the water to fall to the lower level and causing the screw to rotate. This rotational energy can then be extracted by an electrical generator connected to the main shaft of the screw.

Archimedean screws for hydropower are used on low head/high flow sites. They can work efficiently on heads as low as 1 metre, though are not generally used on heads less than 1.5 m (more for economic reasons than technical ones). Single screws can work on heads up to 8 metres, but above this multiple screws are generally used, though in many cases for heads above 8 metres there may be more appropriate turbines available with much smaller footprints.

The maximum flow rate through an Archimedean screw is determined by the screw diameter. The smallest screws are just 1 metre diameter and can pass 250 litres/second, then they increase in 250 mm steps all of the way up to 5 metres in diameter with a maximum flow rate of around 14.5 m3/s. The 5 metre maximum is really based on practical delivery restrictions, and in many cases 3 metres is the maximum diameter that can be delivered to a site. If there is more flow available, multiple screws can be installed in parallel.

In terms of power output, the very smallest Archimedean screws can produce as little as 5 kW, and the largest 500 kW.

The main parts of an Archimedean screw used as a hydro generator are shown below. The actual screw is below the upper bearing. The helical screw or 'flights' are made from rolled flat steel plate that is then welded to a central steel core. Most Archimedean screws have three flights, or three separate helices winding around the central core.

Main parts at the top end of an Archimedean screw hydro turbine generator

Archimedean screws typically rotate at around 26 rpm, so the top of the screw connects to a gearbox to increase the rotational speed to between 750 and 1500 rpm to make it compatible with standard generators. Even though they rotate relatively slowly Archimedean screws can splash water around, though this is reduced significantly by the use of a splash guard shown running down the left-hand side of the screw as shown below.

Archimedean Screw hydro turbine body

Also quite a few Archimedean screws have been installed without any protective guarding over the screw itself, though we would recommend having the whole screw covered to prevent large debris, animals or even people falling in and becoming entrained. The guarding can be designed sympathetically so that the screw is still visible if required.

Archimedean Screw hydro turbine screws are normally set at an angle of 22 degrees from horizontal, which is the optimum for the most cost-effective installations. There is scope to adjusting the angle slightly if the site requires it (to fit into a particular space for example).

The best Archimedean Screw hydro turbine screws are variable-speed in operation, which means that the rotational speed of the screw can be increased or decreased depending on the flow rate available in the river. This is much better than having a fixed-speed screw and varying the flow rate through an automated sluice, which creates high head losses and impacts the overall system efficiency. Variable-speed screws are also quieter in operation and don't suffer from 'back slap' at the discharge-end of the screw.

A typical efficiency curve for a good quality variable-speed Archimedean screw is shown below. This is the mechanical efficiency, so doesn't include the gearbox, generator and inverter losses (these are approximately 15% on in total). It's worth noting that there are some Archimedean screw suppliers that 'over sell' the efficiency of screws, so be careful when comparing performance. A lower claimed efficiency may not be because a particular screw is inferior; it could just be that the supplier is more honest.

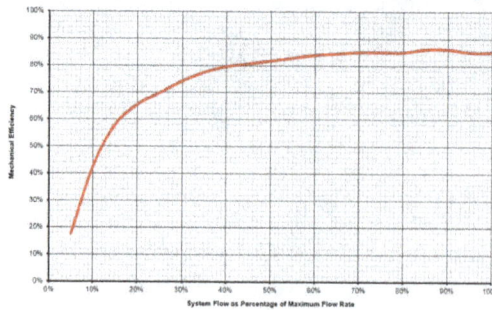

Typical Archimedean Screw hydro turbine efficiency curve

A good quality Archimedean Screw hydro turbine has a design life of 30 years, and this can be extended with a major overhaul which includes re-tipping the screw flights.

A significant advantage of an archimedean screw hydro turbine is their debris tolerance. Due to the relatively large dimensions of the screw's flights and slow rotational speed, relatively large debris can pass through unhindered and without damaging the screw, and certainly all small debris such as leaves can pass through without any problems at all. This means that fine screens are not required at the intake to the screw and they can manage with course screens with 100 or 150 mm bar-spacing. This leads to relatively modest amounts of debris build-up on the course screen and removes the requirement for (expensive) automatic intake screen cleaners which are normally required on larger low-head hydropower systems.

Archimedean-screw-body-awaiting-installation

The low rotational speed and large flow-passage dimensions of the archimedean screw hydro turbine screws also allow fish to pass downstream through the screw in relative safety. Archimedean screws are often touted as 'fish friendly' hydro turbines, which they undoubtedly are, though we at Renewables First would say that all hydro systems should be fish friendly, regardless of turbine type. In non-screw hydro systems this just means well designed intake screens and fish passes / by passes would be required. Note that if upstream fish passage is required at an Archimedean screw hydro turbine site, a fish pass will be required.

The intake side of an Archimedean screw installation

The final advantage of the Archimedean screw hydro turbine is simplified civil engineering works and foundations. Because screws don't have draft tubes or discharge sumps, it means that the depth of any concrete works on the downstream-side of the screw is relatively shallow, which reduces construction costs. The civils works are also relatively simple, the main part being the load-bearing foundations underneath the upper and lower bearings. In softer ground conditions the load-bearing foundations can be piled.

Cross-flow Turbine

Cross-flow turbines, sometimes known as Mitchell-Banki or Ossberger turbines are a type of turbine that tends to be used in smaller hydroelectric sites with power outputs between 5-100 kW. These turbines are useful for a large range of hydraulic heads, from only 1.75 meters to 200 meters, although usually crossflow turbines are chosen for heads below 40 meters.

In addition to being used in smaller hydroelectric facilities, one benefit of these turbines is they require comparatively less complex maintenance to keep them working. Because of this, they are more suitable for use in remote communities.Although useful for a wide range of hydraulic heads and power outputs, generally these turbines are most efficient for low heads and low power outputs. Other turbines are likely more efficient and useful for large-scale applications.

Design

A crossflow turbine is designed using a large cylindrical mechanism composed of a central rotor surrounded by a "cage" of blades arranged into a water wheel shape. These blades are generally sharpened to increase the efficiency of the turbine by reducing the resistance to water flow. Water is directed onto the turbine through a nozzle that creates a flat sheet of water, and then is directed onto the blades using a guide vane. Water first hits the blades and moves to the inside of the turbine, with the water hitting the blades one more time as the water exits the center of the turbine.

Operation

In this type of turbine, water enters as a flat sheet instead of a round jet - as is the case in Pelton turbines. The sheet of water is guided onto the blades of the turbine by an inlet guide vane, ensuring that the water hits the blades at the proper angle to maximize efficiency. The water flows over the blades creating a torque on these blades. After hitting the blades, the sheet of water moves through the turbine and hits the blades once more as it leaves, producing more torque. The first impact the water has with the blades produces more power than the second hit.

Since this type of turbine uses water jets that create an impulse on the turbine, the crossflow turbine is a type of impulse turbine, similar to the Pelton turbine.

Converting Water into Watts

The "cross flow turbine" is very good at converting the kinetic energy of the water into mechanical torque. But to determine the potential energy of the water flowing in a river or stream it is necessary to determine both the flow rate of the water and the head through which the water can be made to fall. The potential power available in the water can be calculated as follows:

Theoretical power: (P) = Flow rate (Q) x Head (H) x Gravity (g)

Where:

> Q = the quantity of flowing water in cubic metres per second (m³/sec)

> H = the effective Head in metres (m)

> g = gravitational acceleration (9.81 m/s²)

We can say that the theoretical power available is: P = 9.81 x Q x H (kW)

However, energy is always lost when it is converted from one form to another in the form of frictional losses and spillages. Although higher than other types of turbine design, cross flow water turbines rarely have efficiencies better than 85%. However, by careful design, this loss can be reduced to only a small percentage but a rough guide used for small systems of a few kW rating is to take the overall efficiency as approximately 50%. Thus, the theoretical power must be multiplied by 0.50 for a more realistic figure.

There is widespread potential around the world for localised hydro energy production from rivers and canals using "zero head" cross flow turbine designs which require no dams or barrages. The kinetic energy in rivers and canals is a predictable energy resource, which is readily available 24 hours a day and can be exploited by using current cross flow turbine designand technology.

Reaction Turbine

A reaction turbine is different from impulse turbine in many ways. In reaction turbine pressure is not remains same throughout the turbine. When water enters into the turbine runner, one part of the available energy of water converts into kinetic energy and remaining part converts into pressure energy.

The pressure varies throughout the turbine. At the entrance of turbine pressure is much higher than the pressure at exit. When water starts flows through the runner the pressure energy starts converts into kinetic energy. Due to variation in pressure the casing of the turbine should be air tight and always full with water. At the entrance pressure is equal to atmospheric pressure but in runner's pressure starts decreases and it is less than atmospheric pressure. This difference in pressure is the reason of flowing of liquids i.e. liquid flow high pressure region to lower pressure region. The difference between the pressures of runner is known as reaction pressure and these types of turbines are known as reaction turbine. Francis, Propeller and Kaplan are the some commonly used reaction turbines.

Operating Conditions

Generally reaction turbines are medium and low head turbines with medium and high discharge.

Reaction Turbine

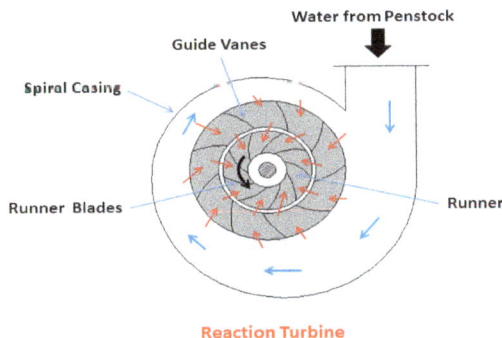

Reaction Turbine

Construction

Reaction turbine has various components which are given as follow:

Reservoir

Reservoir is a large area where water stores. It is continuous source of water with large amount of hydraulic energy. This can be natural as lake or may be artificial called as dam. Large quantity of water store here for continuous supply of water.

Penstock

Penstock is a large diameter pipe which is used to carry the water from reservoir to turbine.

Surge tank

Surge tank is a type of reservoir of water located   near to the turbine which is used to avoid water hammering in penstock.

Casing

Casing is made up of cast steel, plate steel or may be concrete depending upon the working conditions of turbine. It is spiral in shape and gradually decreasing in area. The purpose of the casing is to provide constant velocity of water at the inlet of the runner and to maintain the constant velocity for the water. Gradually decreasing area helps to maintain the constant velocity of water throughout the runner.

Guide Vanes and Fixed Vanes

Fixed vanes have two functions. It guides the water from casing to guide vanes and it also helps in distribution of load due to internal pressure of water. These fixed vanes are generally made up of cast iron, cast steel or fabricated steel.

 The guide vanes use to guide the water towards the runner in guide vane angle direction. These vanes are fixed but can rotate about their own axis. The guide vanes are airfoil in shape and generally made up of cast steel, stainless steel and plate steel. These guides vanes are controlled with the help of governor by controlling flow area to control the discharge or may be operated either by means of wheel in case of small units.

Runner

The runner of reaction turbines generally consists of series of curved vanes mounted circumstantially in the angular space between two plates. These curved vanes are 16 to 24 in number. The runner is generally made up of cast iron, cast steel, or stainless steel. To reduce the cost only some portion of the runner usually made up of stainless steel which subjected to cavitation. The runner is joins with the shaft for power transmission. The torque generated by the runner by the help of hydraulic energy of water is transferred to generator though shaft. This shaft is joins with bolted flange connection.

The runner may be different types as per the requirements. Different turbines have different type of runner because the flow of water is different in all the turbines. Reaction turbines are classifies according to the direction of flow of water with respect to runner. These are axial flow, radial flow and mixed flow type runners. In Kaplan and propeller shaft turbines axial flow runner is used where as in Francis turbine radial flow runner in use. In special case we use mixed flow runner i.e. in modern Francis turbine. So selection of type of runners is depends only on the type of turbines.

Draft tube

Draft tube is the major component in case of reaction turbines, we have no requirement of draft tube in case of impulse turbine. Draft tube is a pipe with gradually increase in cross sectional area. It is fitted at the runner exit to tail race. Draft tube is used to convert the kinetic energy into pressure energy in order to increase the efficiency of the turbine. It is made up of cast steel, plate steel or may be of concrete. Draft tube must be air tight in all conditions and the lower part of the draft tube must be submerged into the water of tail race up to some level. The draft tube used can be of different shape and sizes as per the requirements.

Working

Water is supplied by penstock from reservoir to turbine than enters into the casing. Casing is completely surrounds the runner. This casing distributes the water circumferentially into the runner of turbine. This casing always filled with water. Inside the casing number of fixed vanes present, this converts the head available with water partially into dynamic head. The cross sectional area of the casing gradually decreases to maintain the constant velocity of water throughout into the turbine runner. These various components and vanes help in flowing of water into the runner with minimum loss of energy.

When the water enters over the rotor in the runner it has both kinetic energy and pressure energy. When the water strikes over the moving vanes/ curved vanes it applies impulse force due to kinetic energy same as in case of Pelton wheel. As the water flows over the moving/curved vane it creates a pressure difference across the vane due to air foil shape of the vane, due to which water applies the lift force over the vane. This lift force is also known as reaction force. The impulse and reaction force will rotates the runner. Due to this reason sometimes it is also called as impulse reaction turbine. After runner water is out through the draft tube which is attached at the bottom of the runner. The draft tube provides suction head at the runner exit. The exit water goes into the tail race which further utilize in various applications.

Francis Turbine

Francis turbine is a type of hydraulic turbine which is used to convert hydraulic energy of water into mechanical energy. This mechanical energy is further used for different purpose like electricity generation. Francis turbine is a type of reaction turbine i.e. here pressure is not remain same throughout the turbine. Francis turbine is named on James B. Francis. He developed inward radial flow reaction turbine called as Francis turbine. Later on some modification was made on it then it is called as modern Francis turbine. In modern Francis turbine water enters into the turbine runner radially and leaves axially through its center. Due to this reason sometime it is known as mixed flow reaction turbine.

Francis turbine works on medium head and medium discharge i.e. it works on 30-600 meter head. Now a day's Francis turbine is the most commonly used turbine because of its working conditions. The casing of this turbine should be air tight because pressure varies continuously throughout the runner. At the entrance of the turbine pressure is much higher than the pressure at exit. When water starts flow through the runner the pressure energy start converting into kinetic energy. This turbine use both kinetic energy and pressure energy for working i.e. Francis turbine requires both reaction force and impulse force for power generation.

Construction

Construction of Francis turbine is quite complicated, it has various components which should be arranged in proper manner and should be manufactured with high accuracy. Following are the main components of Francis turbine.

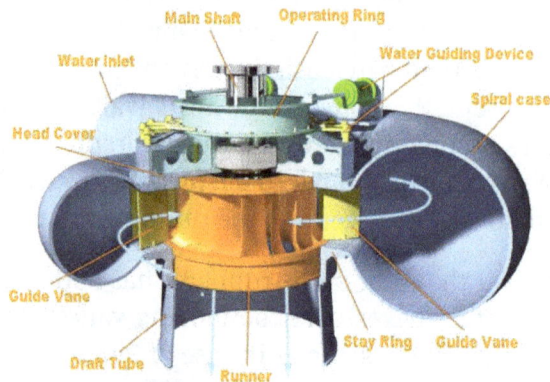

Francis Turbine

Penstock

Penstock transfer water from reservoir to the turbine. It is a large diameter pipe, generally made up of concrete or cast steel.

Spiral Casing

Most of these machines have vertical shafts although some smaller machines of this type have horizontal shaft. The fluid enters from the penstock (pipeline leading to the turbine from the reservoir at high altitude) to a spiral casing which completely surrounds the runner. This casing is known as scroll casing or volute. The cross-sectional area of this casing decreases uniformly along the circumference to keep the fluid velocity constant in magnitude along its path towards the guide vane.

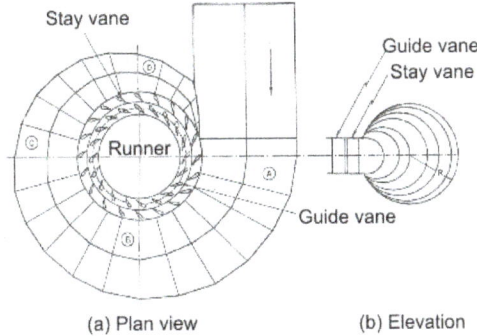

(a) Plan view   (b) Elevation

Figure: Spiral Casing

This is so because the rate of flow along the fluid path in the volute decreases due to continuous entry of the fluid to the runner through the openings of the guide vanes or stay vanes.

Guide or Stay Vane

The basic purpose of the guide vanes or stay vanes is to convert a part of pressure energy of the fluid at its entrance to the kinetic energy and then to direct the fluid on to the runner blades at the angle appropriate to the design. Moreover, the guide vanes are pivoted and can be turned by a suitable governing mechanism to regulate the flow while the load changes. The guide vanes are also known as wicket gates. The guide vanes impart a tangential velocity and hence an angular momentum to the water before its entry to the runner. The flow in the runner of a Francis turbine is not purely radial but a combination of radial and tangential. The flow is inward, i.e. from the periphery towards the centre. The height of the runner depends upon the specific speed. The height increases with the increase in the specific speed. The main direction of flow change as water passes through the runner and is finally turned into the axial direction while entering the draft tube.

Draft Tube

The draft tube is a conduit which connects the runner exit to the tail race where the water is being finally discharged from the turbine. The primary function of the draft tube is to reduce the velocity of the discharged water to minimize the loss of kinetic energy at the outlet. This permits the turbine to be set above the tail water without any appreciable drop of available head. A clear understanding of the function of the draft tube in any reaction turbine, in fact, is very important for the purpose of its design. The purpose of providing a draft tube will be better understood if we carefully study the net available head across a reaction turbine.

Net Head Across a Reaction Turbine and the Purpose to Providing a Draft Tube

The effective head across any turbine is the difference between the head at inlet to the machine and the head at outlet from it. A reaction turbine always runs completely filled with the working fluid. The tube that connects the end of the runner to the tail race is known as a draft tube and should completely to filled with the working fluid flowing through it. The kinetic energy of the fluid finally discharged into the tail race is wasted. A draft tube is made divergent so as to reduce the velocity at outlet to a minimum. Therefore a draft tube is basically a diffuser and should be designed properly with the angle between the walls of the tube to be limited to about 8 degree so as to prevent the flow separation from the wall and to reduce accordingly the loss of energy in the tube. Figure 28.3 shows a flow diagram from the reservoir via a reaction turbine to the tail race.

The total head H_1 at the entrance to the turbine can be found out by applying the Bernoulli's equation between the free surface of the reservoir and the inlet to the turbine as,

$$H_o = \frac{P_1}{\rho g} + \frac{V_1^2}{2g} + z + h_f$$

or,

$$H_1 = H_o - h_f = \frac{P_1}{\rho g} + \frac{V_1^2}{2g} + z$$

Where h_f is the head lost due to friction in the pipeline connecting the reservoir and the turbine. Since the draft tube is a part of the turbine, the net head across the turbine, for the conversion of mechanical work, is the difference of total head at inlet to the machine and the total head at discharge from the draft tube at tail race and is shown as H in figure:

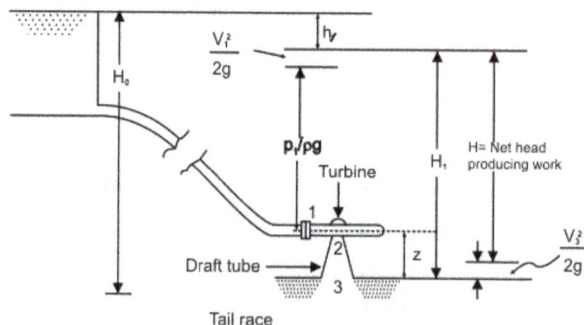

Figure: Head across a reaction turbine

Therefore,

H = total head at inlet to machine- total head at discharge,

$$\frac{P_1}{\rho g} + \frac{V_1^2}{2g} + z - \frac{V_3^2}{2g} = H_1 - \frac{V_3^2}{2g}$$

$$= (H_0 - h_f) - \frac{V_3^2}{2g}$$

The pressures are defined in terms of their values above the atmospheric pressure. If the losses in the draft tube are neglected, then the total head at becomes equal. Therefore, the net head across the machine is either $(H_1 - H_3)$ or $(H_1 - H_2)$.

Applying the Bernoull's equation,

$$\frac{P_2}{\rho g} + \frac{V_2^2}{2g} + z = 0 + \frac{V_3^2}{2g} +$$

$$\frac{P_2}{\rho g} = -\left[z + \frac{V_2^2 - V_3^2}{2g}\right]$$

Since $V_3 < V_2$, both the terms in the bracket are positive and hence $P_2/\rho g$ is always negative, which implies that the static pressure at the outlet of the runner is always below the atmospheric pressure. Equation $H_0 = \frac{P_1}{\rho g} + \frac{V_1^2}{2g} + z + h_f$ also shows that the value of the suction pressure at runner outlet depends on z, the height of the runner above the tail race and $(V_2^2 - V_3^2)/2g$, the decrease in kinetic energy of the fluid in the draft tube. The value of this minimum pressure p_2 should never fall below the vapour pressure of the liquid at its operating temperature to avoid the problem of cavitation. Therefore, we fine that the incorporation of a draft tube allows the turbine runner to be set above the tail race without any drop of available head by maintaining a vacuum pressure at the outlet of the runner.

Runner of the Francis Turbine

The shape of the blades of a Francis runner is complex. The exact shape depends on its specific speed. It is obvious from the equation of specific speed that higher specific speed means lower head. This requires that the runner should admit a comparatively large quantity of water for a given power output and at the same time the velocity of discharge at runner outlet should be small to avoid cavitation. In a purely radial flow runner, as developed by James B. Francis, the bulk flow is in the radial direction. To be more clear, the flow is tangential and radial at the inlet but is entirely radial with a negligible tangential component at the outlet. The flow, under the situation, has to make a $90°$ turn after passing through the rotor for its inlet to the draft tube. Since the flow area (area perpendicular to the radial direction) is small, there is a limit to the capacity of this type of runner in keeping a low exit velocity. This leads to the design of a mixed flow runner where water is turned from a radial to an axial direction in the rotor itself. At the outlet of this type of runner, the flow is mostly axial with negligible radial and tangential components. Because of a large discharge area (area perpendicular to the axial direction), this type of runner can pass a large amount of water with a low exit velocity from the runner. The blades for a reaction turbine are always so shaped that the tangential or whirling component of velocity at the outlet becomes zero $\left(V_{w_2} = 0\right)$. This is made to keep the kinetic energy at outlet a minimum.

Figure just below shows the velocity triangles at inlet and outlet of a typical blade of a Francis turbine. Usually the flow velocity (velocity perpendicular to the tangential direction) remains constant throughout, i.e. $V_{f_1} = V_{f_2}$ and is equal to that at the inlet to the draft tube.

The Euler's equation for turbine reduces,

$$E / m = e = V_{w1} U_1$$

where, e is the energy transfer to the rotor per unit mass of the fluid. From the inlet velocity triangle shown infigure below.

$$V_{w1} = V_{f_1} \cot \alpha_1$$

$$U_1 = V_{f_1} \left(\cot \alpha_1 + \cot \beta_1 \right)$$

and

Substituting the values of V_{w1} and U_1 from equation $V_{w1} = V_{f_1} \cot \alpha_1$ and $U_1 = V_{f_1} \left(\cot \alpha_1 + \cot \beta_1 \right)$ respectively into equation. ($E / m = e = V_{w1} U_1$), we have,

$$e = V_{f_1}^2 \cot \alpha_1 \left(\cot \alpha_1 + \cot \beta_1 \right)$$

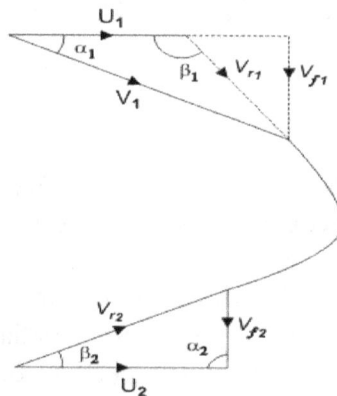

Figure: Velocity triangle for a Francis runner

The loss of kinetic energy per unit mass becomes equal to $V_{f_1}^2 / 2$. Therefore neglecting friction, the blade efficiency becomes,

$$\eta_b = \frac{e}{e + \left(V_{f_2}^2 / 2\right)}$$

$$= \frac{2V_{f_1}^2 \cot \alpha_1 \left(\cot \alpha_1 + \cot \beta_1\right)}{V_{f_2}^2 + 2V_{f_1}^2 \cot \alpha_1 \left(\cot \alpha_1 + \cot \beta_1\right)}$$

Since,

$$V_{f_1} = V_{f_2} . \eta b$$

can be written as,

$$\eta_b = 1 - \frac{1}{1 + 2 \cot \alpha_1 \left(\cot \alpha_1 + \cot \beta_1\right)}$$

The change in pressure energy of the fluid in the rotor can be found out by subtracting the change in its kinetic energy from the total energy released. Therefore, we can write for the degree of reaction.

$$R = \frac{e - \frac{1}{2}\left(V_1^2 - V_{f_2}^2\right)}{e} = 1 - \frac{\frac{1}{2}V_{f_1}^2 \cot^2 \alpha_1}{e}$$

[since $V_1^2 - V_{f_2}^2 = V_1^2 - V_{f_1}^2 = V_{f_1}^2 \cot^2 \alpha_1$]

Using the expression of e from equation ($e = V_{f_1}^2 \cot \alpha_1 \left(\cot \alpha_1 + \cot \beta_1\right)$),

we have,

$$R = 1 - \frac{\cot \alpha}{2\left(\cot \alpha_1 + \cot \beta_1\right)}$$

The inlet blade angle β_1 of a Francis runner varies $45 - 120°$ and the guide vane angle angle α_1 from $10 - 40°$. The ratio of blade width to the diameter of runner B/D, at blade inlet, depends upon the required specific speed and varies from $1/20$ to $2/3$.

Expression for specific speed. The dimensional specific speed of a turbine, can be written as,

$$N_{sT} = \frac{NP^{1/2}}{H^{5/4}}$$

Power generated P for a turbine can be expressed in terms of available head H and hydraulic efficiency η_h as,

$$P = \rho \, Q \, g \, H \, \eta_h$$

Hence, it becomes,

$$N_{sT} = N \left(\rho \, Q \, g \, \eta_h \right)^{1/2} H^{-3/4}$$

Again, $N = U_1 / \pi D_1$,

Substituting U_1 from equation $U_1 = V_{f_1} \left(\cot \alpha_1 + \cot \beta_1 \right)$

$$N = \frac{V_{f_1} \left(\cot \alpha_1 + \cot \beta_1 \right)}{\pi D_1}$$

Available head H equals the head delivered by the turbine plus the head lost at the exit. Thus,

$$g H = e + \left(V_{f_2}^2 / 2 \right)$$

since,

$$V_{f_1} = V_{f2}$$

$$g H = e + \left(V_{f_1}^2 / 2 \right)$$

with the help of equation $e = V_{f_1}^2 \cot \alpha_1 \left(\cot \alpha_1 + \cot \beta_1 \right)$ it becomes,

$$g H = V_{f_1}^2 \cot \alpha_1 \left(\cot \alpha_1 + \cot \beta_1 \right) + \frac{V_{f_1}^2}{2}$$

or,

$$H = \frac{V_{f_1}^2}{2g} \left[1 + 2 \cot \alpha_1 \left(\cot \alpha_1 + \cot \beta_1 \right) \right] \tag{29.7}$$

Substituting the values of H and N from equations $H = \frac{V_{f_1}^2}{2g} \left[1 + 2 \cot \alpha_1 \left(\cot \alpha_1 + \cot \beta_1 \right) \right]$ and $N = \frac{V_{f_1} \left(\cot \alpha_1 + \cot \beta_1 \right)}{\pi D_1}$ respectively into the expression N_{sT} given by equation

$$N_{sT} = N \left(\rho \, Q \, g \, \eta_h \right)^{1/2} H^{-3/4}$$

we get,

$$N_{sT} = 2^{3/4}\, g^{5/4} \left(\rho\, \eta_h\, Q\right)^{1/2} \frac{V_{f_1}^{-1/2}}{\pi D_1}\left(\cot\alpha_1 + \cot\beta_1\right) + \left[1 + 2\cot\alpha_1\left(\cot\alpha_1 + \cot\beta_1\right)\right]^{-3/4}$$

Flow velocity at inlet V_{f_1} can be substituted from the equation of continuity as,

$$V_{f_1} = \frac{Q}{\pi D_1 B}$$

where B is the width of the runner at its inlet.

Finally, the expression for N_{sT} becomes,

$$N_{sT} = 2^{3/4}\, g^{5/4} \left(\rho\, \eta_h\,\right)^{1/2} \left(\frac{B}{\pi D_1}\right)^{1/2} \left(\cot\alpha_1 + \cot\beta_1\right)$$

$$\left[1 + 2\cot\alpha_1\left(\cot\alpha_1 + \cot\beta_1\right)\right]^{-3/4}$$

For a Francis turbine, the variations of geometrical parameters like $\alpha_1, \beta_1 B/D$ have been described earlier. These variations cover a range of specific speed between 50 and 400. Figure below shows an overview of a Francis Turbine. The figure is specifically shown in order to convey the size and relative dimensions of a typical Francis Turbine to the readers.

Figure: Installation of a Francis Turbine

Working

Working of Francis turbine quite complicated. Water enters into the turbine through penstock then enters into the spiral casing. The cross sectional area of casing decrease circumferentially to maintain constant velocity of water. After casing water enters into the runner, here water first strikes to fixed blades then guide vanes. Fixed vanes convert the head available with water partially into dynamic head. The casing of the Francis turbine should be air tight and always filled with water. Fixed vanes remove swirls from the water and make linear flow of the water. Swirls generates due to the spiral casing. After that water strikes to guide vanes which are stationary at their place but revolves around their own axis. Guide vanes decides the proper angle at which water strikes to the runner blades and control the flow rate of water into the runner. The guide vanes are controlled by means of wheel in case of small units or by using some automatically operated governor in case of larger units. When load is increases the flow rate of water increases by means of guide vanes and as the load decreases, flow rate automatically decreases. All these components are designed to lead the water into the runner with minimum loss of energy.

The runner of Francis turbine have series of curved vanes, these vanes are designed so that water enters into the runner radially and leaves it in axial direction in case of modern Francis turbine but in case of old version simple Francis turbine water in and out both in radially. This depends on the construction of the runner. This change in the direction of runner from radial to axial produces a circumferential force on the runner and runner starts rotating and output is taken by joining shaft with runner. Shaft is coupled with the runner which is used for power transmission. With the help of shaft power is transferred to the generator which is used for generation of electricity.

Applications

Francis turbine can be operated for different head this is the main advantage of this turbine and due to this Francis turbine has various applications:

1. It is most commonly used turbine in hydro power plant for electricity generation.

2. Mixed flow reaction turbine has some other applications in irrigation purposes like pump the water from the ground and many more.

3. It is the most efficient turbine these days, so it can be used for different purposes for power generation by using hydraulic energy of water.

Propeller Turbine

The Propeller Turbine is an inward flow reaction turbine, similar to a Kaplan design, but with fixed blades. It is a very common turbine and works best with high flow rates. Its moving part (runner) is a propeller, similar to those that push ships and submarines through water.

The turbine has adjustable guide vanes that control the water flow in the turbine. They also direct the water at an angle to the back of the propeller.

Pico Propeller Turbine

The pico propeller turbine uses a closed scroll casing that directs water onto the propeller. No guide vanes are needed, and the propeller blades are fixed. Tests show that guide vanes tend to cause energy losses in this type of turbine. The water exits through a conical draft tube.

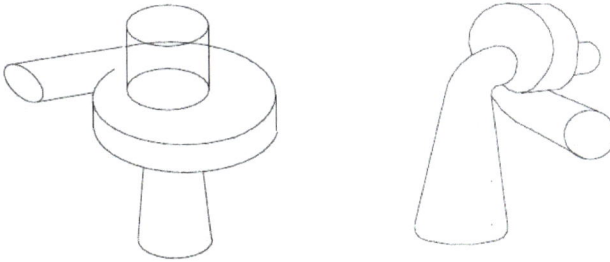

Figure: Possible Turbine layouts (shown with a direct-drive generator)
(a) vertical shaft (b) horizontal shaft

Other layouts are possible for small axial-flow turbines but this particular layout was chosen for the following reasons:

- Standard bearings can be used, as the runner is "overhung";

- It is possible to have a direct drive to the generator, without need for a long shaft;

- The design can be used for a wide range of heads (without likelihood of cavitation);

- The turbine can be manufactured using basic mechanical workshop equipment.

Turbine Design Spreadsheet

The design procedure follows a series of logical steps. The design starts from the runner and then calculates the dimensions of the scroll casing to give the correct swirl velocity at runner inlet. The dimensions of the runner may need to be adjusted through several iterations to avoid large twist in the blades or reverse angles at the hub, which would make the runner blades difficult to manufacture accurately. Choice of materials and manufacturing methods are left to the knowledge of the designer.

Turbine Design Parameters

The key design parameters for a turbine are head (H), volume flow, or discharge (Q) and rotational speed (N). These values can be put into the spreadsheet "Sizing" page. From these three parameters, a "dimensionless shape number" or "specific speed" can be determined. This number gives an indication of the geometry of the turbine and it is the starting point for detailed design.

There are many different forms of the specific speed. For this design procedure we use the following equation:

$$n_q = \frac{N\sqrt{Q}}{H^{0.75}}$$

Where, N is in rev/min, Q in m³/s and H in m.

So, the starting point for the turbine design is to decide the values of Q, N and H. One of the key findings of the authors was that measuring the available flow rate accurately is critical for effective design of the turbine, as there is no adjustment for flow variation when the turbine has been made. If the actual flow available is lower than the turbine design flow, Q, the turbine will generate very little power.

The choice of the speed, N, depends on the speed of the generator and the type of drive used. Often it is possible to use a direct drive, with the turbine runner attached to the end of an extended generator shaft. On the other hand, using a single stage belt drive allows the possibility of changing the turbine operating speed. This gives more flexibility in the turbine design and in matching to site conditions.

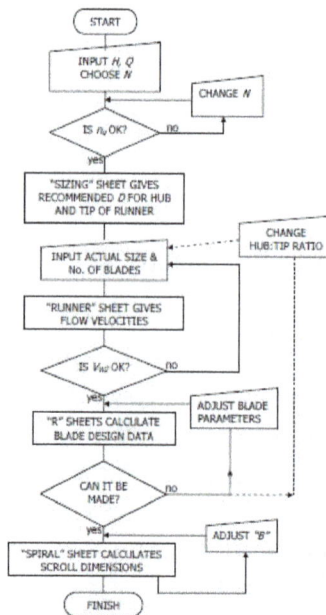

Figure: Flow Chart for the Design Spreadsheet

Specific Speed and Varying Flow

The expected range of specific speed values is 70 < nq < 300. If the specific speed is less than 70, then you should look at an alternative type of turbine – e.g. crossflow (Mitchell-Banki), pump as turbine or turgo.

You should avoid designing a turbine with a specific speed greater than 250 as this will tend to have a low efficiency. Where head is low and flow high, it might be a good idea to design for parallel turbines, each operating on part of the total flow. Otherwise it is necessary to choose a lower speed for the turbine, which will result in a large physical size. Two smaller turbines running at higher speed may be not much more costly than one large turbine.

If the turbine flow varies between two main seasons in the year, then it may be possible to design two different runners that can be changed over, one for higher flow and one for lower flow. Both runners can be designed to operate with good efficiency in the same casing, but additional care is needed in the design, as they will have different specific speeds. It is planned to give an example in a later version of this manual.

Key Design Choices

There are some decisions about the turbine that the designer needs to make during the design process. For example, there are two options for the spiral casing design – a parallel sided casing or a tapered casing. For the runner blades there is a choice between aerofoil cross-section and constant thickness blades. There is also a choice to be made regarding the material for the runner.

Shaft Orientation

The choice between horizontal and vertical shaft will depend on the application and the designer's preference. This decision is important as it will affect the power house layout. A horizontal shaft layout of the turbine has the following advantages:

- Allows a mechanical load to be connected more easily;

- Easier to change rotors if the need arises;

A vertical shaft arrangement may reduce losses in the draft tube because it does not need a 90-degree bend. It may also have advantages if using a direct-drive generator:

- Any water leakage is likely to run away from the generator;

- There is less axial load on the extended shaft.

Scroll Casing Design

There is a significant difference in the flow pattern produced by a constant height spiral casing as compared with a tapered spiral design; this affects the matching of the scroll casing and the runner blade design. The advantage of the tapered casing is that it is easier to create a suitable flow angle into the runner and less material is used for the same size of turbine, making it slightly lighter to transport. However, for very low heads (below 3 - 4 m) the tapered scroll is not recommended because it creates a high velocity at the inlet to the runner, which causes a very low pressure to occur. This leads to two potential problems: i) the shaft seal tends to suck in air; ii) there may be low enough pressure for cavitation to occur. Both of these problems are serious as they lead to a reduction in turbine efficiency and may also cause damage to the turbine runner.

Blade Design

The choice of aerofoil blades or constant thickness blades relates to the manufacturing process. Tests on prototype turbines showed that good quality of manufacture of the blades is important to obtain high efficiency – each blade needs to match the design. Good surface finish also improves efficiency. However, achieving the required blade twist (the change of angle from hub to tip) is more important than having a complex blade profile.

Aerofoil blades would normally be made by casting individual blades that are then welded onto the hub. These could be of aluminium or bronze. Constant thickness blades would normally be used with a steel runner. They can be made out of sheet steel, cut, bent and twisted into shape and then welded onto the hub. When welding the blades, it is recommended to use a jig to hold each blade at the correct angle to the hub. It is important that the runner blades are evenly spaced around the runner and are each set at the same angle.

Guide to "Sizing" Sheet

The value of the tip velocity to the head velocity is a key parameter in the turbine sizing. The equation for this ratio is:

$$k_{ug} = \frac{r_{tip} \times \omega}{\sqrt{2gH}}$$

Where rtip is the blade tip radius, i.e. rtip = D/2 and ω is the angular velocity of the turbine runner, in rad/s, i.e. $\omega = 2\pi N/60$. In the spreadsheet kug is calculated according to the specific speed, based on the graph shown below.

There are two other parameters shown on this graph, which are input by the designer in the "Sizing" sheet. These are the diameter ratio of runner hub:tip and the number of runner blades (Z). The values on the graph are for guidance. The values for the number of blades have been modified because for small turbines it is better to have fewer blades on the runner. This graph is adapted from a book by Willi Bohl that is itself based on efficient Kaplan turbine designs.

Runner Design Parameters

When choosing the hub:tip ratio, remember that a smaller ratio will mean a relatively large twist needed in the blade. For ease of manufacture of small high specific speed turbines, it may be advisable to use an even larger value than indicated by the range on the graph.

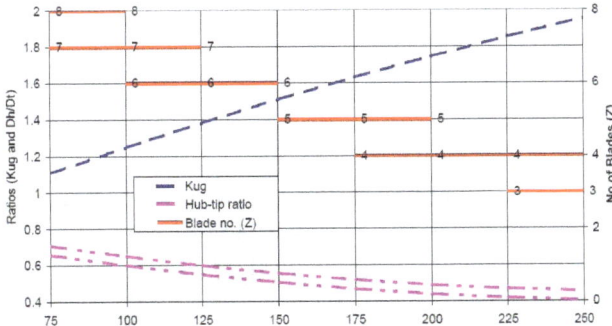

Figure Design Parameters for "Sizing" sheet

Example of Turbine Options

As an example of the options produced by the "Sizing" sheet, table gives the dimensions for turbines to fit a head of 2.5 m and total flow of 460 litre/s.

	1 turbine	2 (direct drive)	2 (belt drive)
Turbine Flow (l/s)	460	230	230
Speed (rpm)	720	1030	720
Specific Speed (nq)	246	248	174
Runner dia. (mm)	360	250	300

To achieve the recommended nq < 250 either the flow is limited or the speed must be reduced. 1030 rpm has been chosen as it is suitable for direct drive of a standard 6-pole 50 Hz induction motor as generator. The detailed design in the example spreadsheet has been carried out for the middle option.

Hydraulic Efficiency

An estimate of the hydraulic efficiency is required for calculating the turbine dimensions. At the bottom of the sizing sheet there are two estimates of the hydraulic efficiency. These are based on statistical data on a large number of turbines, collated by Anderson.

The upper figure shows the mean hydraulic efficiency for turbines with this flow rate. The lower figure takes into account the specific speed. It demonstrates the efficiency disadvantage of choosing a high specific speed machine. This value, or an estimate from elsewhere, can be input at the top of the spreadsheet.

Runner Design Guide

The process of designing the runner using *constant thickness* blades is outlined in the flowchart on the next page. This worksheet picks up data from the "Sizing" worksheet giving the basic dimensions of the runner. It calculates the shapes of the blades required when using constant thickness, circular arc blades. It is later intended to publish a similar sheet for designing profiled (aerofoil) blades.

According to basic theory, using the Euler equation:

$$g\eta_h H = u_1 v_{w1} - u_2 v_{w2}$$

Where u is the peripheral speed of the runner (1 at inlet and 2 at outlet) and vw is the tangential (or whirl) component of the water velocity. For an axial flow turbine $u_1 = u_2$. The axial flow velocity through the runner is calculated.

$$\text{from:} \quad v_a = \frac{Q}{\pi r_t^2 - r_h^2} = \frac{Q}{\pi D_t^2 - D_h^2}$$

Where r_t is the radius of the outside of the runner (tip radius) and rh is the radius of the inside edge of the runner blade (hub radius).

In the spreadsheet, the value of v_{w1r} is input as a variable for the designer to choose. In some cases – e.g. where a new runner is being designed for an existing casing – this is already fixed by the scroll design, but normally it can be adjusted, thus changing the values of vw2 and w2. An initial guess for vw1r is given in the spreadsheet, based on the Euler Equation and assuming (as a first guess) that vw2 is positive and is 10% of v_{w1}.

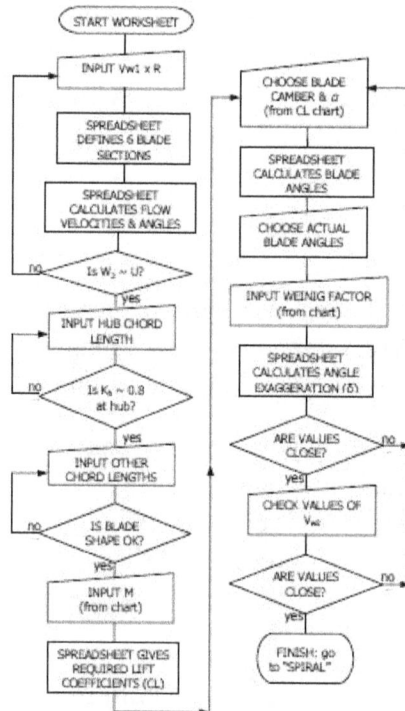

Figure: Flowchart for "Runner" worksheet

Exit Velocity Triangle

For a simple analysis, it is often assumed that there is no whirl velocity at exit, $vw2 = 0$. However, the angles calculated from this method are different from the actual blade angles in efficient turbines, and a more complicated procedure is recommended. There are several reasons for this:

- The flow does not follow the blades exactly, because there is a relatively large space between each blade;

- The blade of an axial turbine acts as an aerofoil, and operates better with a positive "angle of attack" in order to produce a good torque on the shaft;

- The optimum energy transfer does not necessarily occur when exit whirl velocity is zero.

Nechleba, for example, recommends that the outlet velocity triangle has equal sides ($w2 = u$) on the basis that this gives highest overall efficiency. For some designs this gives a significant value of $vw2$. A value of $vw2$ somewhere between that for which $w2 = u$ and $vw2 = 0$ can then be used. The spreadsheet calculates the flow angles from the velocity triangles and the Euler equation. Note that an adjustment is made to take account of additional losses at the blade hub and tip.

$$w_1^2 = \alpha - V_{w1} + V_a^2; \quad w_2^2 = \alpha - V_{w2} + V_a^2;$$

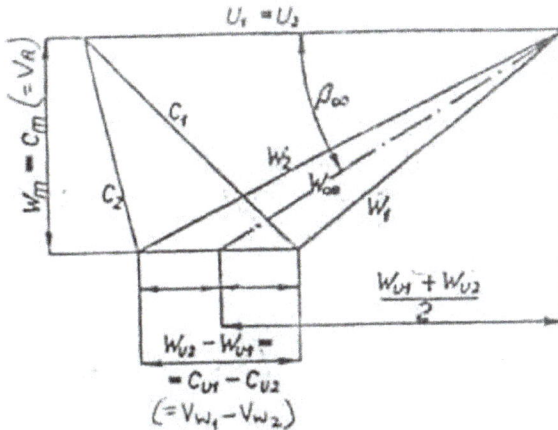

Figure: Diagram of flow velocities, with w2 = u.

Note that, since $u = \omega r$ and $v_{w1} \propto 1/r$, (free vortex flow from the scroll casing) v_{w2} will also be approximately proportional to $1/r$, so decreases from hub to tip.

The flow angles are calculated from:

$$\tan \beta_1 = \frac{V_a}{u - V_{w1}}; \quad \tan \beta_2 = \frac{V_a}{u - V_{w2}}; \quad \beta_\infty = \frac{\beta_1 + \beta_2}{2};$$

Runner Blade Length

Various recommendations are given in the literature for the length of the blade "chord", normally as a ratio of the blade pitch (x), which varies with the radius from hub to tip.

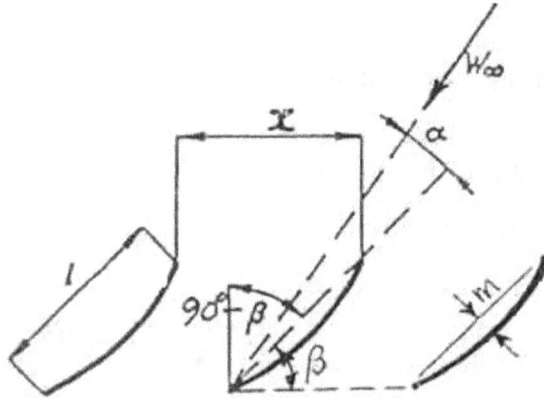

Figure: Definition of angles and blade dimensions

Usually the chord length l should increase from hub to tip, but the ratio of pitch:chord, x/l, should also increase from hub to tip. In the spreadsheet, the chart "blade shape" can be used to check that the blade is not a strange shape to manufacture. In the "Co-efficients" spreadsheet there is a chart "space chord ratios" that shows a number of recommendations for the values of x/l varying from hub to tip and as a function of the turbine specific speed. These come from Wu, Ytreøy and Raabe.

From the chord lengths, the "Runner" spreadsheet calculates a factor, k_b, called "blade loading factor" which is used to find the required lift coefficients.

It is given in Nechleba's book by the equation:

$$k_b = \frac{2gH\eta_h \, {}^x\!/_l}{uw_\infty} = MC_L - \frac{C_D}{\tan \beta_\infty} + \frac{C_L^2}{6\pi \tan \beta_\infty}$$

However, the last two terms are found to be insignificant, especially given the lack of accuracy in estimating M and hence CL. It is suggested that the value of kb at the hub should be approximately 0.8. This is known as the Zweifel criterion, which is ironic, since Zweifel, in German, means "doubt" and there is some doubt as to the applicability of this criterion for the case of water turbines, since it was originally developed for compressible fluids.

The value of M depends on the value of β_∞ and x/l, according to the chart below. The "Runner" spreadsheet now calculates the required values of lift coefficient, CL, for each of the blade sections.

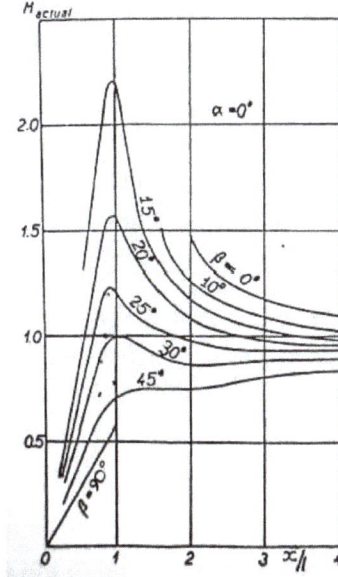

Figure: Factor to modify Lift Coefficients due to multiple blades (adapted from Nechleba). Although it is strictly for α = 0, it is applicable to other values, as long as the blades are relatively thin.

Runner Blade Shape

The blade shape has three variables: the angle of attack, α; the thickness, t; the camber, m, in this case given as a percentage of chord length, l. In the spreadsheet "Coefficients" there is a chart of lift coefficients for thin, constant thickness blades. This is based on values from fan blade tests, but the Reynolds number is similar to that in water turbines, so the results should be applicable. Normally, the angle of attack and the blade camber will both increase from blade tip to blade hub. The values should be chosen to give the values of lift coefficient in the spreadsheet line CL.

The next rows on the "Runner" spreadsheet lead to the calculation of the actual blade setting angle (beta calculated), by subtracting the angle of attack. It also calculates the axial length of the blade (Lhub). This is particularly important at the hub section, because it determines the minimum axial length of the runner hub. The angles can be rounded off or adjusted and are input manually, after which the spreadsheet calculates the solidity – the amount of each circumferential section that is covered by blade when looking parallel to the turbine axis. In case these values look impractical, it may be necessary to go back and adjust the blade chord length or even the number of blades.

The next two rows in the spreadsheet calculate the arc radius and deflection angle, based on a circular arc. There is now an additional check that can be carried out to complete the design. Again, as used by fan designers (who are most used to applying constant thickness curved blades), there is a relationship between the flow deflection angle ($\beta_1 - \beta_2$) and the blade deflection angle. The flow is deviated less than the blade deflection, and there is an "angle exaggeration factor", δ, that can be used to find the ratio of flow deflection to blade deflection, based on the work of Weinig as quoted by Bohl and Bommes. The chart in spreadsheet "Coefficients" gives the value of δ as a function of β_∞ and x/l.

If the actual value of "theta ratio" is greater than the value extracted from the chart, then the CL chart should be used again. The angle of attack should be reduced and the camber increased to give the same required value of lift coefficient, but with a greater value of blade deflection.

As a final check, the spreadsheet re-calculates the exit flow angle, based on the angle exaggeration and then re-calculates the exit whirl velocity, v_{w2}. The value of exit whirl should agree fairly closely with the values highlighted in blue earlier in the spreadsheet.

Casing Design Guide

This guide and the current "Spiral" worksheet calculate the dimensions for a spiral with a tapering rectangular cross-section. There are three parameters that have to be input at the top of the worksheet: firstly, the radius at the inner edge of the spiral (where most larger turbines have guide vanes – hence Rg). This must be larger than the runner radius in order to allow for a chamfer or radius at the turn into the runner tube section. A radius (as shown on the right hand side in figure below) is preferable to a chamfer, but may be more difficult to manufacture. In either case, allow an increase of at least 5% of the runner diameter. The second parameter is the height of the inlet section at the inner radius of the spiral, before the flow changes direction into the runner, b_0. Finally, the cross-section of the spiral inlet, B is input by the designer.

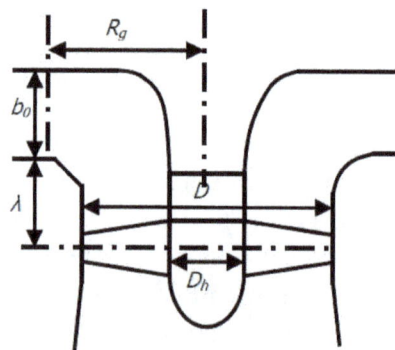

Figure: Dimensions of Runner Inlet

It is recommended that $0.35 \le b_0/D \le 0.5$, increasing with increasing specific speed. The value of B is then adjusted so that the two values of "$Q/V_w R_w$" are as close as possible, which is when:

$$\frac{Q}{V_w R} = B \log_e \left(1 + \frac{B}{R_g} \right)$$

The height of the spiral is also calculated and a chart shows the how the height decreases with spiral angle. This can be used to check that there is not a large discontinuity towards the centre of the spiral. The value of the runner setting height, λ, must also be decided. Nechleba and Bohl both recommend about D/4.

Draft Tube Guidelines

A draft tube recovers some of the kinetic energy from the runner outlet and should reduce losses due to non-uniform flow into the tailrace. A simple conical draft tube is normally acceptable, but in some cases a 90° bend is required to (e.g. for a horizontal axis turbine). Any bend must be designed with parallel or increasing diameter, so that it does not restrict the outflow from the turbine. A divergence angle for draft tube of between 8° and 12° is recommended, i.e. for a symmetrical vertical cone, the sides should be 4° to 6° from the vertical. The draft tube length should be between 4 and 10 times the diameter of the runner, giving an outlet diameter of 1.8 to 2.5 times the runner diameter. Note that the axial velocity through the draft tube will be reduced in proportion to the area – i.e. by a factor of between approximately 3 and 6.

Bulb Turbine

The bulb turbine is a reaction turbine of Kaplan type which is used for extremely low heads. The characteristic feature of this turbine is that the turbine components as well as the generator are housed inside a bulb, from which the name is developed. The main difference from the Kaplan turbine is that the water flows in a mixed axial-radial direction into the guide vane cascade and not through a scroll casing. The giude vane spindles are normally inclined to 60° in relation to the turbine shaft and thus results in a conical guide vane cascade contrary to other types of turbines. The runner of a bulb turbine may have different numbers of blades depending on the head and water flow. The bulb turbines have higher full-load efficiency and higher flow capacity as compared to Kaplan turbine. It has a relatively lower construction cost. The bulb turbines can be utilized to tap electrical power from the fast flowing rivers on the hills. Figure below shows the schematic of a Bulb Turbine Power Plant.

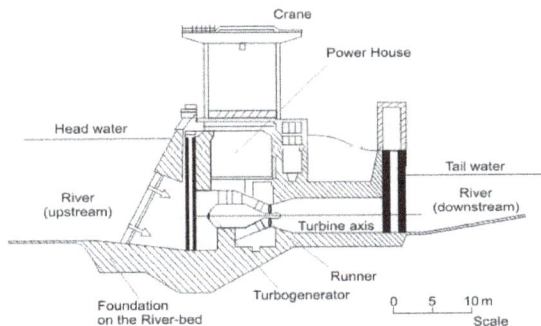

Figure: Schematic of Bulb Turbine Power Generating Station

Characteristics

While the bulb turbine is the most common solution for high outputs at low headsites, S- and pit turbines are frequently favored for economic solutions in small hydro applications with outputs up to about 10 MW. Specific project requirements determine, which hydroelectric equipment is favorable on a case by case basis.

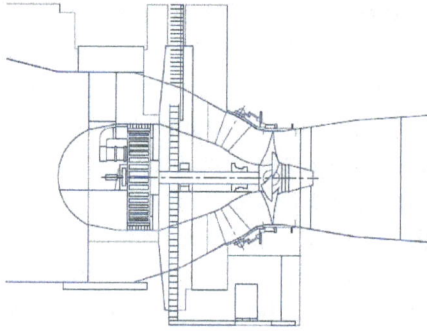

Cross section of a bulb turbine and generat   Cross section of a pit turbine, gearbox and generator

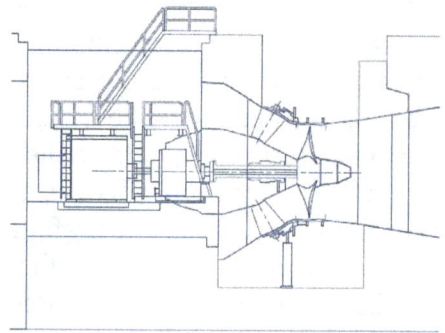

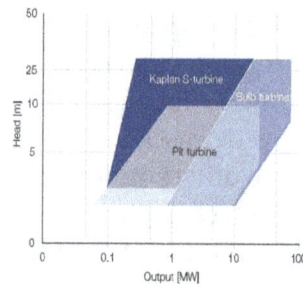

The application of pit and S-turbine units provides unique advantages. Their design provides good accessibility of various components and assures reliability and long service life.

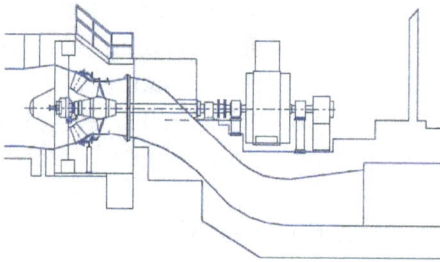

Cross section of an S-turbine and generator   Application range

Advantages of the Bulb Design

Higher full-load efficiency and higher flow capacities of bulb and pit turbines can offer many advantages over vertical Kaplan turbines.

In the overall assessment of a low head project, the application of bulb/pit turbines results in higher annual energy and lower relative construction costs.

Pit-type turbines with a speed increaser located between the runner and generator are used for projects with heads lower than 10 meters.

Generator Characteristics of Bulb

As bulb units grew in size, generator voltage also grew from initially 3.3 and 6.6 kV to 11.0 kV or even 13.8 kV on higher capacity units. Cooling systems evolved from originally separate heat exchangers to maintenance-free closed loop systems, providing heat dissipation directly into the river water passing the bulb unit. For very high capacities and high speed units pressurized air can also be used to improve heat dissipation.

Most bulb units are arranged with two bearings: a guide bearing near the overhung turbine runner, and a combined guide and thrust bearing supported by the stay column just downstream of the generator.

The bearing systems of horizontal machines, whether for bulbs, pit-turbines or S-type machines, are arranged to handle the counter thrust associated with load rejections of such units.

Trend of bulb generator output

Three-bearing systems are provided for certain high capacity machines, optimized for specific project requirements. The challenge in modern bulb unit designs is to achieve maximum reliability and availability, with minimum maintenance of the main units and associated auxiliaries.

3D section of a vertical bulb unit

With its broad experience in hydro generators for the world's largest hydroelectric fa-cilities, and fifty years of experience in the design, manufacture and installation of bulb generators, Voith can provide fully optimized bulb, pit and S-type units for any instal-lation.

Straflo

The Straflo axial flow turbine is high specific speed fully tubular machine and used for low head and high discharge. Straflo turbines comprises turbine and generator into single unit are the best choice for micro hydro power development from water flowing in pipes. The generator is mounted around the periphery of runner. The blades of axial flow turbines are highly twisted due to which there occurs change in angular momen-tum which forces the rotor to rotate along with generator shaft which in turn generates electricity.

The Straflo turbine is designed for a discharge of 550 kg/s, a head of 7.5 m and a pow-er output of 35 KW. The axi-symmetric axial distributor with fixed 7 vanes, propeller runner with 4 fixed blades and conical draft tube are designed. The computed values of inlet and outlet blade angles are used in Bladegen to generate blades. A suitable stagger angle variation is taken from hub to shroud so as to get blades with smooth curvature. The inlet flow angle for runner is taken as blade outlet angle for guide vanes. Axial distributor with twisted blades is used ahead of the runner. The three components are modelled separately with only a single blade for both the runner and distributor. Tetra/mixed mesh using robust (octree) method is used for meshing in Ansys ICEM CFD for whole of the turbine space separately for all components. The simulations have been carried out for different mesh sizes to check mesh dependency of flow parameters and to get Y+ value with acceptable limits. The distributor, runner and draft tube are mod-elled and meshed separately and connected through proper interfaces. The geometry of the turbine model used for simulations is shown in figure below:

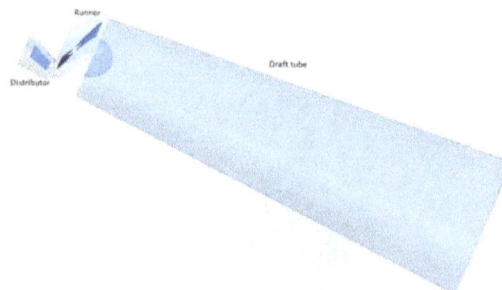

Figure: 3D geometry of straflo turbine model for flow simulations

Boundary Conditions

The flow simulations have been carried out for three different mass flow rates specified at inlet of distributor and three different rotational speed of runner, including design and off design conditions. Smooth walls with no slip are chosen as boundary condition for blades, hub, shroud, draft tube cone. Reference pressure is taken as zero atmosphere. Periodicity is applied at runner and distributor because of the limited computational power. Distributor and draft tube are taken as stationary domain and runner as rotating domain. Frozen rotor type of interface is used for interfacing between stationary and rotating domains. The SST turbulence model is used in simulation for the flow inside turbine being complex and rotating.

Formulae Used

The characteristic parameters are computed in dimensionless form using following formulae:

Specific energy coefficient $\psi = \dfrac{g\,HD^*}{Q^2}$

Speed factor $SF = \dfrac{ND}{\sqrt{gH}}$

Discharge factor $DF = \dfrac{Q}{D^2\sqrt{gH}}$

Runner head $H_R = \dfrac{(P_{01} - P_{02})_{S\tan\ Frame} - (P_{01} - P_{02})_{Rot.frame}}{\rho g}$

Hydraulic efficiency (%) $\eta_h = \dfrac{H_R}{H} * 100$

Pressure coefficient $C_p = \dfrac{P - P_2}{\rho \dfrac{w_2^2}{2}}$

Hydraulic losses (%) $H_L = \dfrac{(H - H_R)}{H} * 100$

Degree of reaction $R = \dfrac{W_2^2 - W_1^2}{2gH}$

Draft Turbine

In power turbines like reaction turbines, Kaplan turbines, or Francis turbines, a diffuser tube is installed at the exit of the turbine, known as draft tube.

This draft tube at the end of the turbine increases the pressure of the exiting fluid at the expense of its velocity. This means that the turbine can reduce pressure to a higher extent without fear of back flow from the tail race.

In an impulse turbine the available head is high and there is no significant effect on the efficiency if the turbine is placed a couple of meters above the tail race. But in the case of reaction turbines, if the net head is low and if the turbine is installed above the tail race, there can be appreciable loss in available pressure head to power the turbine. Also, if the pressure of the fluid in the tail race is higher than at the exit of the turbine, a back flow of liquid into the turbine can result in significant damage.

By placing a draft tube (also called a diffuser tube or pipe) at the exit of the turbine, the turbine pressure head is increased by decreasing the exit velocity, and both the overall efficiency and the output of the turbine can be improved. The draft tube works by converting some of the kinetic energy at the exit of the turbine runner into the useful pressure energy.

Using a draft tube also has the advantages of placing the turbine structure above the tail race so that any required inspections can be made more easily and reducing the amount of excavation required for construction.

Efficiency

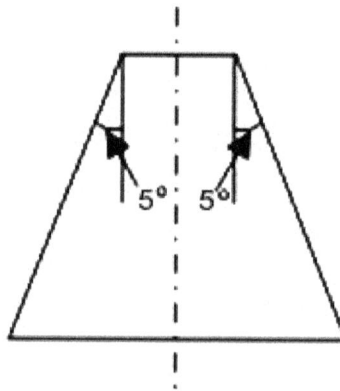

Conical Diffuser

Conical draft tube

It is defined as the ratio of the actual conversion of kinetic energy into pressure energy in the draft tube to the kinetic energy available at the draft tube inlet.

η= Difference of kinetic energy between inlet and outlet-tube losses/Kinetic Energy at the inlet.

$$\eta_{dt} = \frac{(V_2^2 - V_3^2) - 2gh_d}{V_2^2}$$

V_2 = Fluids velocity at inlet of draft tube or at the outlet of turbine

V_3 = Fluids velocity at outlet of draft tube

g= gravitational acceleration

h_d = head losses in draft tube

Draft Tube allows turbine to be placed above the tail race and simultaneously allows it to operate at the same efficiency if it was placed at the tail race.

Draft Tube and Cavitation

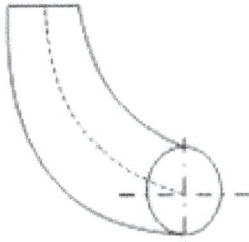

Simple elbow type

simple elbow draft tube

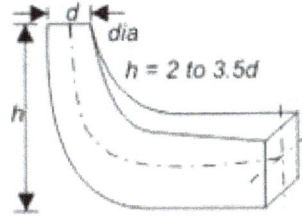

Elbow type with varying cross-section

Elbow type with rectangular cross section

Cavitation occurs when the local absolute pressure falls below the saturated vapor pressure of the water for the water temperature. The height of draft tube is an important parameter for avoiding cavitation. Applying Bernoulli's equation between outlet of the runner and discharge point of the draft tube neglecting any head losses in draft tube),

$$z_2 + \frac{p_2}{\rho g} + \frac{V_2^2}{2g} = z_3 + \frac{p_3}{\rho g} + \frac{V_3^2}{2g}$$

Where,

z_2 = z (Height of draft tube)

z_3 = height of tail race which is referenced as datum line (=0)

p_2 = pressure at the outlet of the runner

p_3 = gauge pressure

$$\frac{p_2}{\rho g} = -\left[z + \frac{V_2^2 - V_3^2}{2g} \right]$$

Since draft tube is a diffuser V_3 is always less than V_2 which implies p_2 is always negative thus height of the draft tube is an important parameter to avoid cavitation.

Types of Draft Tube

1. Conical diffuser or straight divergent tube: This type of draft tube consists of a conical diffuser with half angle generally less than equal to $10°$ to prevent flow separation. It is usually employed for low specific speed,vertical shaft francis turbine. Efficiency of this type of draft tube is 90% ;

2. Simple elbow type draft Tube: It consists of an extended elbow type tube. Generally, used when turbine has to be placed close to the tail-race. It helps to cut down the cost of excavation and the exit diameter should be as large as possible to recover kinetic energy at the outlet of runner. Efficiency of this kind of draft tube is less almost 60% ;

3. Elbow with varying cross section: It is similar to the Bent Draft tube except the bent part is of varying cross section with rectangular outlet.the horizontal portion of draft tube is generally inclined upwards to prevent entry of air from the exit end.

Tyson Turbine

The Tyson Turbine is a hydropower system that extracts power from the flow of water. This design doesn't need a casement, as it is inserted directly into flowing water. It consists of a propeller mounted below a raft, driving a power system, typically a generator, on top of the raft by belt or gear. The turbine is towed into the middle of a river or stream, where the flow is the fastest, and tied off to shore. It requires no local engineering, and can easily be moved to other locations. The Tyson Turbine is a very common way to reuse energy.

Gorlov Helical Turbine

The Gorlov helical turbine (GHT) is a water turbine evolved from the Darrieus turbine design by altering it to have helical blades/foils. It was patented in a series of patents from September 19, 1995 to July 3, 2001 and won 2001 ASME Thomas A. Edison Patent Award. GHT was invented by Professor Alexander M. Gorlov of Northeastern University.

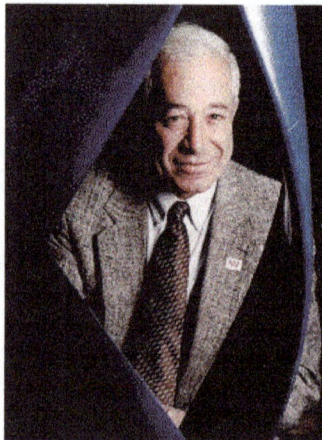

A.M.Gorlov with his turbine.

The physical principles of the GHT work are the same as for its main prototype, the Darrieus turbine, and for the family of similar Vertical axis wind turbines which includes also Turby wind turbine, aerotecture turbine, Quietrevolution wind turbine, GHT, turby and quietrevolution solved pulsatory torque issues by using the helical twist of the blades.

Fluid Performance

The term "foil" is used to describe the shape of the blade cross-section at a given point, with no distinction for the type of fluid, (thus referring to either an "airfoil" or "hydrofoil"). In the helical design, the blades curve around the axis, which has the effect of evenly distributing the foil sections throughout the rotation cycle, so there is always a foil section at every possible angle of attack. In this way, the sum of the lift and drag forces on each blade do not change abruptly with rotation angle. The turbine generates a smoother torque curve, so there is much less vibration and noise than in the Darrieus design. It also minimizes peak stresses in the structure and materials, and facilitates self-starting of the turbine. In testing environments the GHT has been observed to have up to 35% efficiency in energy capture reported by several groups. "Among the other vertical-axis turbine systems, the Davis Hydro turbine, the EnCurrent turbine, and the Gorlov Helical turbine have all undergone scale testing at laboratory or sea. Overall, these technologies represent the current norm of tidal current development."

Turbine Axis-orientation

The main difference between the Gorlov helical turbine and conventional turbines is the orientation of the axis in relation to current flow. The GHT is a vertical-axis turbine which means the axis is positioned perpendicular to current flow, whereas traditional turbines are horizontal-axis turbines which means the axis is positioned parallel to the flow of the current. Fluid flows, such as wind, will naturally change direction, however they will still remain parallel to the ground. So in all vertical-axis turbines, the flow remains perpendicular to the axis, regardless of the flow direction, and the turbines always rotate in the same direction. This is one of the main advantages of vertical-axis turbines.

If the direction of the water flow is fixed, then the Gorlov turbine axis could be vertical or horizontal, the only requirement is orthogonality to the flow.

Airfoil/Hydrofoil

The GHT operates under a lift-based concept. The foil sections on the GHT are symmetrical, both top-to-bottom and also from the leading-to-trailing edge. The GHT can actually spin equally well in either direction. The GHT works under the same principle as the Darrieus turbine; that is, it relies upon the movement of the foils in order to change the apparent direction of the flow relative to the foils, and thus change the (apparent) "angle of attack" of the foil.

Environmental Issues

A GHT is proposed for low-head micro hydro installations, when construction of a dam is undesirable. The GHT is an example of damless hydro technology. The technology may potentially offer cost and environmental benefits over dam-based micro-hydro systems.

Some advantages of damless hydro are that it eliminates the potential for failure of a dam, which improves public safety. It also eliminates the initial cost of dam engineering, construction and maintenance, reduces the environmental and ecological complications, and potentially simplifies the regulatory issues put into law specifically to mitigate the problems with dams.

In general, a major ecological issue with hydropower installations is their actual and perceived risk to aquatic life. It is claimed that a GHT spins slowly enough that fish can see it soon enough to swim around it. From preliminary tests in 2001, it was claimed that if a fish swims between the slowly moving turbine blades, the fish will not be harmed. Also it would be difficult for a fish to become lodged or stuck in the turbine, because the open spaces between the blades are larger than even the largest fish living in a small river. A fish also would not be tumbled around in a vortex, because the GHT does not create a lot of turbulence, so small objects would be harmlessly swept through with the current.

Working of Gorlov Helical Turbine

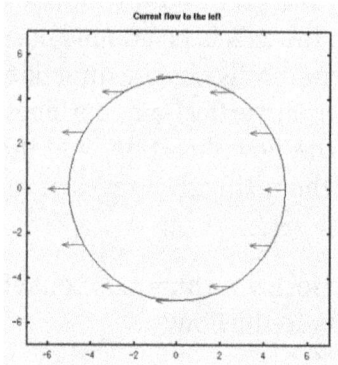

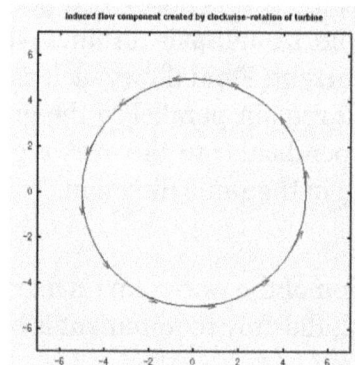

a. Current flow to the left   b. Induced flow component created by clockwise rotation of turbine

In this example the direction of the fluid flow is to the left. As the turbine rotates, in this case in a clockwise direction, the motion of the foil through the fluid changes the apparent velocity and angle of attack (speed and direction) of the fluid with respect to the frame of reference of the foil. The combined effect of these two flow components (i.e. the vector sum), yields the net total "Apparent flow velocity" as shown in the next figure.

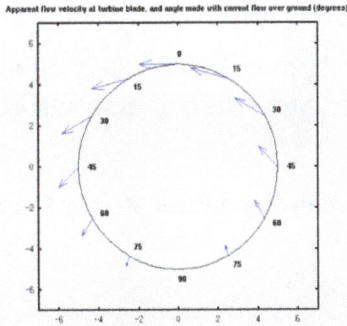

Apparent flow velocity of turbine blade, and angle
made with current flow over ground (in degrees).

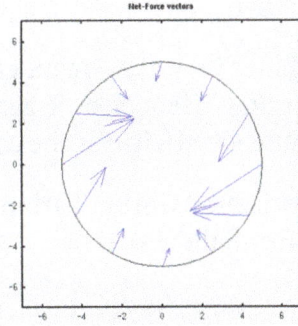

Net force vectors.

The action of this apparent flow on each foil section generates both a lift and drag force, the sum of which is shown in the figure above titled "Net force vectors". Each of these net force vectors can be split into two orthogonal vectors: a radial component and a tangential component, shown here as "Normal force" and "Axial force" respectively. The normal forces are opposed by the rigidity of the turbine structure and do not impart any rotational force or energy to the turbine. The remaining force component propels the turbine in the clockwise direction, and it is from this torque that energy can be harvested.

With regards to the figure above left "Apparent flow velocity", Lucid Energy Technologies, rights holder to the patent to the Gorlov Helical Turbine, notes that this diagram, with no apparent velocity at an azimuth angle of 180 degrees (blade at its point in rotation where it is instantaneously moving in downstream direction), may be subject to misinterpretation. This is because a zero apparently flow velocity could occur only at a tip speed ratio of unity (i.e. TSR=1, where the current flow induced by rotation equals the current flow). The GHT generally operates at a TSR substantially greater than unity.

The diagrams "Net Force Vectors" and "Normal Force Vectors" are partially incorrect. The downwind segments should show the vectors outside the circles. Otherwise there would be no net sideways loading on the turbine.

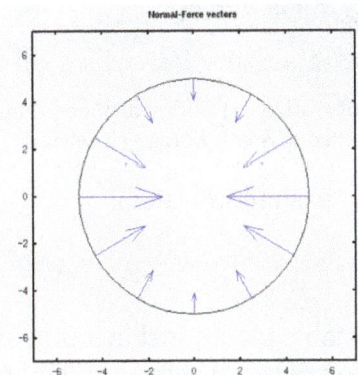

Normal force vectors.

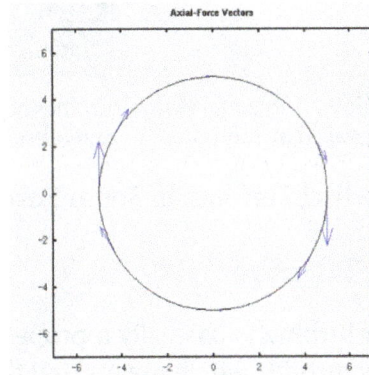

Axial force vectors.

Commercial Use

Helical turbines in water stream generate mechanical power independent on direction of the water flow. Then electric generators assembled upon the common shaft transfer the power into electricity for the commercial use.

Chain of Horizontal Gorlov Turbines. TideGen by Ocean Renewable Power Company – possibility for shallow waters.

Chain of Horizontal Gorlov turbines

Tidal Power Station with Gorlov Helical Turbines before deployment in the ocean.

Chain of Horizontal Gorlov turbines being installed in South Korea - general view

Chain of Horizontal Gorlov turbines being installed in South Korea - close view

Gorlov Helical Turbines in South Korea. Installation in shallow waters.

Kaplan Turbine

A Kaplan turbine is basically a propeller with adjustable blades inside a tube. It is an axial-flow turbine, which means that the flow direction does not change as it crosses the rotor. Figure shows a simplified Kaplan turbine.

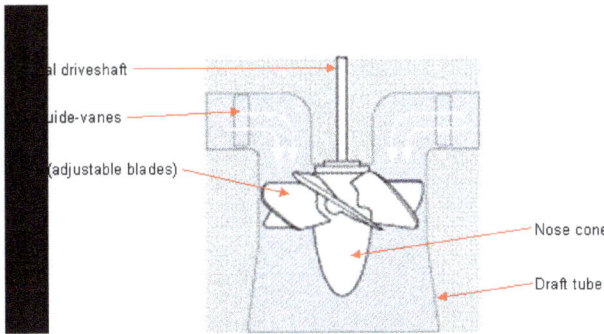

Figure: Basic layout of a Kaplan turbine

The inlet guide-vanes can be opened and closed to regulate the amount of flow that can pass through the turbine. When fully closed they will stop the water completely and bring the turbine to rest. Depending on the position of the inlet guide-vanes they introduce differing amounts of 'swirl' to the flow, and ensure that the water hits the rotor at the most efficient angle for the highest efficiency. The rotor blade pitch is also adjustable, from a flat profile for very low flows to a heavily-pitched profile for high flows. This adjustability of both inlet guide-vanes and rotor blades means that the flow operating range is very wide (a characteristic from the inlet guide-vanes) and the turbine efficiency is high and the efficiency curve very flat (a characteristic from the adjustable rotor blades allowing optimum alignment of the blade to the oncoming flow).

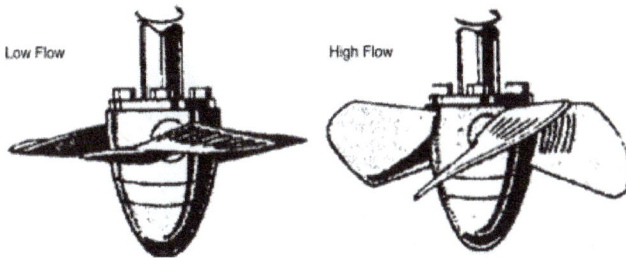

Figure: Kaplan turbine rotor blade positions

The nose cone on a Kaplan turbine is important hydro-dynamically to reduce losses and prevent the formation of a core 'rope vortex', and also provides the space for the complex blade pitching mechanism inside. The draft tube is also a critically important part. Although a static fabricated part, the geometry of the draft tube is carefully designed to extract any remaining kinetic energy from the flow by reducing the water pressure after the rotor.

There are variants of Kaplan turbines that only have adjustable inlet guide-vanes or adjustable rotor blades, which are known as semi-Kaplan's. Although the performance of semi-Kaplan's is compromised when operating across a wide flow range, for applications where the flow does not vary much they can be a more cost-effective choice. Figure below shows how the efficiency varies across the operating flow range for a full-Kaplan (curve A), a semi-Kaplan with adjustable blades (curve B) and a semi-Kaplan with adjustable inlet guide-vanes (curve D). It also shows the efficiency curve for a propeller turbine (a Kaplan with fixed blades and fixed inlet guide-vanes (curve C).

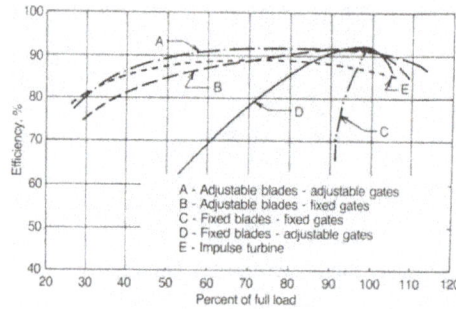

Figure: Kaplan turbine efficiency curve comparison

Kaplan turbines could technically work across a wide range of heads and flow rates, but because of other turbine types being more effective on higher heads, and because Kaplan's are relative expensive, they are the turbine of choice for lower head sites with high flow rates. Typically they are used on sites with net heads from 1.5 to 20 metres and peak flow rates from 3 m³/s to 30 m³/s. In the UK this tends to be on lowland rivers with low heads (1.5 to 5 metres) and relatively high flow rates (up to 30 m³/s). Such systems would have power outputs ranging from 75 kW up to 1 MW.

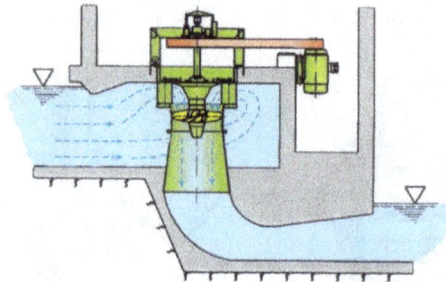

Figure: Cross-section of a typical vertical-axis Kaplan turbine system

The smallest good quality Kaplan turbines available have rotor diameters of 600 mm, though these tend to be prohibitively expensive, at least a very low heads, so generally speaking the smallest rotors are 800 mm. The largest rotors available have 3 to 5 metre diameters. For even larger sites multiple-turbines tend to be used rather than increasing the diameter further. Kaplan turbines are available in three basic configurations; vertical axis, horizontal axis (also called S-turbines) and bulb turbines.

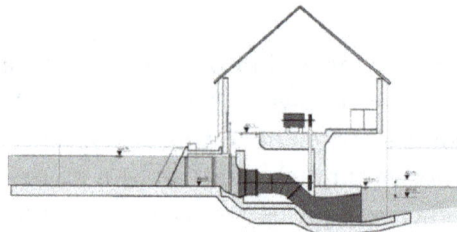

Figure: Cross-section of a typical horizontal axis S-turbine Kaplan turbine system

The commonest orientation currently being installed in the UK (at least by Renewables First) is vertical axis. Vertical-axis Kaplan's have the advantage of requiring the smallest footprint or land-take. A typical layout is shown in figure above. The Kaplan turbine is built into the concrete structure, with the inlet volute (basically a snail shell shaped pipe that wraps around the inlet guide-vanes and distributes the water equally around the whole circumference) and draft tube cast into the concrete at the construction phase. So critical is the perfect geometry of the intake volute and draft tube that it is normal practice for the turbine manufacturer to supply the wooden formwork for these parts to be used by the civil engineering contractor.

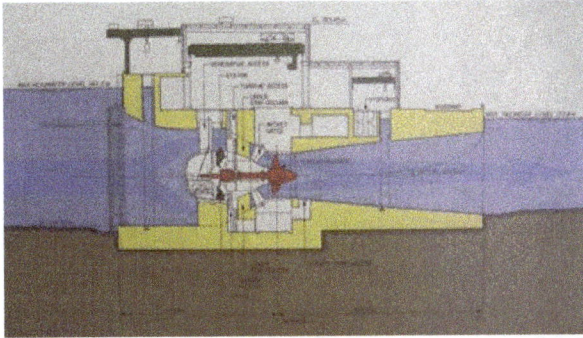

Figure: Cross-section of a typical bulb Kaplan turbine system

Horizontal axis or 'S-turbines' and bulb turbines are technically slightly more efficient than vertical axis Kaplans because the inlet flow does not have to change direction so should have lower hydraulic losses. In reality there is no discernible difference, so the decision on the orientation is normally made through choice of supplier and price. S-turbines do require a larger system footprint which can be a disadvantage in space-constrained sites. A typical S-turbine layout is shown in figure above.

Bulb Kaplan turbines have all of the drive system and generator accommodated inside a streamlined 'bulb' that sits within the main flow. They are only practical on large hydro projects where it is physically possible for a person to climb down into the bulb for maintenance and are normally used on large systems only. A typical cross-section is shown in figure above.

Figure: Ossberger horizontal-axis Kaplan turbine being lifted into position

Figure above shows an HSI horizontal-axis Kaplan turbine with a 1 metre diameter rotor being lifted into position. Figure below shows an S-turbine in operation. If you would like to work with Renewables First on a Kaplan turbine hydro project, please get in touch.

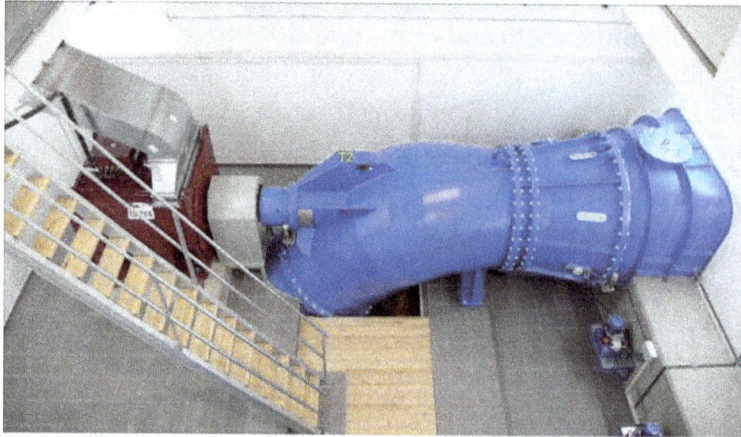

Figure: Kaplan S-turbine in operation

Pros and Cons Pros

Pros:

- Kaplan Turbines can achieve efficiency of up to 95%.

- Can be implemented in low-head situations allowing for power plants at lower elevation.

- Varying size of turbine and output allows for micro-hydropower plants instead of large dams.

- Kaplan Turbines are relatively low cost due to the small size and low head requirements.

- Dams build with the Kaplan Turbine design produces less environmental impact because of low head requirements therefore the reservoir area would not be flooded as much as high head dams.

Cons:

- Often times, Kaplan Turbines are installed where fish migrations occur, and may potentially affect their migration patterns and survival rates.

- The high velocity of the turbine may cause leakage of oil-based lubricant into the outlet leading to pollution.

- Due to the high discharge of the turbine, the water around the turbine can reach very low pressure making the Kaplan turbine vulnerable to cavitation.

References

- Anyi, Martin; Brian Kirke (2010). "Evaluation of small axial flow hydrokinetic turbines for remote communities". Energy for Sustainable Development. 14: 110–116. doi:10.1016/j.esd.2010.02.003

- Hydraulic-turbine-working-types-advantages-and-disadvantages: mech4study.com, Retrieved 16 April 2018

- "The Kinne Collection of Water Turbines" (PDF). The American Society of Mechanical Engineers. 1999-06-19. Retrieved 2009-11-25.

- Waterwheel-engineering, technology: britannica.com, Retrieved 19 May 2018

- Boon, G. C. and Williams, C. The Dolaucothi Drainage Wheel, Journal of Roman Studies, 56 (1966), 122-127

- Impulse-turbine: mechanicalbooster.com, Retrieved 30 March 2018

- Cross-flow-turbine: alternative-energy-tutorials.com, Retrieved 26 June 2018

- Petkewich Rachel, Technology Solutions: Creating electricity with undammed hydropower, Environ. Sci. Technol., 2004, 38 (3), pp 55A–56A, doi:10.1021/es0403716

- Reaction-turbine-basic-principle-construction-and-working: mech4study.com, Retrieved 18 July 2018

- "Rebirth on the River". Pennsylvania Gazette. University of Pennsylvania. January–February 2000. Retrieved 2012-08-16

4

Watermills

A watermill is a structure which employs a water wheel or turbine to drive mechanical processes. This chapter has been carefully written to provide an easy understating of the different kinds of watermills such as tide mills, sawmills, gristmills, etc. and their role in hydropower generation.

A watermill is an engine that uses a water wheel or turbine to drive a mechanical process such as flour or lumber production, or metal shaping (rolling, grinding or wire drawing). A watermill that only generates electricity is more usually called a hydroelectric plant.

Operation of a Watermill

Watermills in Bosnia

Typically, water is diverted from a river or impoundment or mill pond to a turbine or water wheel, along a channel or pipe (variously known as a flume, head race, mill race, leat, leet, lade (Scots) or penstock). The force of the water's movement drives the blades of a wheel or turbine, which in turn rotates an axle that drives the mill's other machinery. Water leaving the wheel or turbine is drained through a tail race, but this channel may also be the head race of yet another wheel, turbine or mill. The passage of water is controlled by sluice gates that allow maintenance and some measure of flood control; large mill complexes may have dozens of sluices controlling complicated interconnected races that feed multiple buildings and industrial processes.

Watermills can be divided into two kinds, one with a horizontal waterwheel on a vertical axle, and the other with a vertical wheel on a horizontal axle. The oldest of these were horizontal mills in which the force of the water, striking a simple paddle wheel set horizontally in line with the flow turned a runner stone balanced on the rind which is atop a shaft leading directly up from the wheel. The bed stone does not turn. The problem with this type of mill arose from the lack of gearing; the speed of the water directly set the maximum speed of the runner stone which, in turn, set the rate of milling.

Types of Watermills

- Gristmills grind grains into flour. These were undoubtedly the most common kind of mill.

- Fulling mills or Walkmills were used for a finishing process on cloth (see also fulling).

- Blade mills were used for sharpening newly made blades.

- Sawmills cut timber into lumber.

- Barking mills stripped bark from trees for use in tanneries.

- Spoke mills turned lumber into spokes for carriage wheels.

- At the beginning of the industrial revolution, cotton mills were usually powered by a water wheel.

- Carpet mills for making rugs were sometimes water-powered.

- Textile mills for weaving cloth were sometimes water-powered.

- Powder mills for making black powder or smokeless powder were sometimes water-powered.

- Blast Furnaces, finery forges, slitting mills, and tinplate works were until the introduction of the steam engine invariably water powered and were sometimes called iron mills.

- Prior to the introduction of the cupola (a reverberatory furnace), lead was usually smelted in smelt mills.

- Paper mills used water not only for motive power but also in large quantities in the manufacturing process.

Paddle Wheel

Paddlewheel flowmeters use the mechanical energy of the fluid to rotate a paddlewheel (just like a riverboat) in the flow stream. Paddles on the rotor are inserted into the flow to transform energy from the flow stream into rotational energy. The rotor shaft spins on bearings. When the fluid moves faster, the paddlewheel spins proportionally faster. Shaft rotation can be sensed mechanically or by detecting the movement of the paddles. Paddle movement is often detected magnetically, with each paddle or embedded piece of metal generating a pulse. When the fluid moves faster, more pulses are generated. The transmitter processes the pulse signal to determine the flow of the fluid.

Paddlewheel flowmeters measure the velocity of liquids in pipes, such as chemicals, water and liquids. High accuracy is attainable if carefully installed. These flowmeters are measuring flow at the edge of the flow profile and thus are affected by viscosity changes. The most common use is in a system where the fluid is like water and other variables such as pH/ORP, conductivity, pressure, temperature and level are monitored. All sensors are inserted into the same pipe Ts and connect into one controller/transmitter. There are temperature and pressure limitations of this insertion system but it is very versatile.

Paddle wheel flowmeters work best with clean fluids—particles can prevent the paddle from spinning properly. They offer fast response time and are easy to maintain. The flowmeters have three components:

- Pipe fitting: available in various operating flow ranges to meet the needs of different applications.

- Paddlewheel sensor: designed to be inserted into the pipe fitting. The sensor consists of the paddlewheel with imbedded magnets and the electronic sensor. About half of the paddle extends into the flow stream. Fluid flowing through the pipe instigates the spinning of the paddlewheel. Magnets in the paddle spin past the sensor. The electrical pulses produced are proportional to the rate of flow.

- Flow display/controller: picks up the signal from the sensor, converts it into an actual flow rate or flow total value, and displays the values. The processed signal can open and close valves, start and stop pumps, indicate high or low flow rate alarms in the system, and more.

Application Cautions for Paddlewheel Flowmeters

Paddlewheel flowmeters are less accurate at low flow rates due to rotor/bearing drag that slows the rotor. Make sure to operate these flowmeters above approximately 5 percent of maximum flow. Paddlewheel flowmeters should not be operated at high velocity because premature bearing wear and/or damage can occur. Be careful when measuring fluids that are non-lubricating because bearing wear can cause the flowmeter become inaccurate and fail. In some applications, sensor replacement may need to be performed routinely and increase maintenance costs. Applications in dirty fluids are not recommended. Easy replacement with insertion magmeters is possible when high failure rates are unacceptable and are often sold by the same vendor. In summary, paddlewheel flowmeters have moving parts that are subject to degradation with time and use.

Abrupt transitions from gas flow to liquid flow should be avoided because they can mechanically stress the flowmeter, degrade accuracy, and/or damage the flowmeter. These conditions generally occur when filling the pipe and under slug flow conditions. Two-phase flow conditions can also cause flowmeters to measure inaccurately.

Tide Mill

A tide mill is quite simply a water mill that derives its power from the rise and fall of the tides. It is almost never referred to as a "seawater" or "saltwater" mill because the chemical composition of the water driving the mill wheel is not important. What counts is that the water impounded behind a mill dam can only be put to work after the water level outside of the dam has sufficiently dropped during the ebb tide. it comprises the building which houses the milling machinery, the mill-pond, where the tidal water is retained, and the dam or causeway which confine and controls it.

Possibly the earliest tide mill in the Roman world was located in London on the River Fleet, dating back to Roman times.

Three Mills, Stratford, one of the world's earliest recorded tide mills.

In recent years, a number of new archaeological finds has consecutively pushed back the date of the earliest tide mills, all of which were discovered on the Irish coast: A 6th century vertical-wheeled tide mill was located at Killoteran near Waterford. A twin flume horizontal-wheeled tide mill dating to c. 630 was excavated on Little Island. Alongside it, another tide mill was found which was powered by a vertical undershot wheel. The Nendrum Monastery mill from 787 was situated on an island in Strangford Lough in Northern Ireland. Its millstones are 830mm in diameter and the horizontal wheel is estimated to have developed 7/8HP at its peak. Remains of an earlier mill dated at 619 were also found at the site.

Three Mills, House Mill and Miller's House at low tide

The earliest *recorded* tide mills in England are listed in the Domesday Book. Eight mills are recorded on the River Lea (the site at Three Mills remains, with Grade I listed buildings and a small museum), as well as a mill in Dover harbour. By the 18th Century there were about 76 tide mills in London, including two on London Bridge.

Woodbridge Tide Mill, an excellent example, survives at Woodbridge, Suffolk, England. This mill, dating from 1170 and reconstructed in 1792, has been preserved and is open to the public. It was further restored in 2010 and re-opened in 2011 in full working order, and became the second working tide mill in the United Kingdom regularly producing flour. Carew Castle in Wales also has an intact, but unused, tide mill. The first tide mill to be restored to working order is Eling Tide Mill in Eling, Hampshire. Another example, now only visible in historic documents, is the mill in the hamlet of Tide Mills, East Sussex. Traces of a tide mill may be seen at Fife Ness, the site of an archaeological survey.

Tidal mill at l'île de Bréhat

A mediæval tide mill still operates at Rupelmonde near Antwerp, and there are several still in existence in the Netherlands.

At one time there were 750 tide mills operating along the shores of the Atlantic Ocean: approximately 300 in North America, including many in colonial Boston over 150 years, 200 in the British Isles, and 100 in France. The Rance estuary in France was also home to some of these mills.

By the mid 20th Century the use of water mills had declined dramatically. In 1938, an investigation by Rex Wailes discovered that of the 23 extant tidal mills in England, only 10 were still working by their own motive power. Of one of the few remaining by the 1940s, at Beaulieu, H. J. Massingham wrote, "Part of the mill is built on piles into the river and is weather boarded, while the rest of the building is a warm red brick roofed with lozenge-shaped and rounded tiles which I believe are called fish-tiles. All the interior is of wood - ladders, bins for the meal, floor-boarding, square pillars, beams, narrow passages, fittings, shaft rising to the first floor and all. So ramshackle is the arrangement of the props and supports that it is a wonder that the whole edifice does not tumble about the miller's ears like a pack of cards. The point is that it has stood in this way for something like six centuries, and that gives the explorer into its dusky depths a more penetrating notion of how the old builders could build, more than does a Gothic church or even a cathedral. The pulse and swing of the great wheel sets the whole building in an ague, but it will still be standing when all the flimsy excrescences of development between Beaulieu and Poole have fallen down."

Newer types of tidal power often propose a dam across a large river estuary. Although it represents a source of renewable energy, each proposal tends to come under local opposition because of its likely impact on coastal habitats. One proposal, which came to fruition in 1966, is the Rance barrage which generates 250MW. Unlike historical tide mills which could only operate on an ebb tide, the Rance barrage can generate electricity on both flows of the tide or it can be used for pumped storage depending on demand. A less intrusive design is for a 1MW free standing turbine, constructed in 2007 at Strangford Lough Narrows - also close to an old tide mill.

Whatever the dam, pond and waterwheel configuration of any particular mill complex, the means of obtaining and releasing water was fairly universal. The power of the incoming tide would force open one or more pairs of tide gates (or valve flaps) to admit the flow of water into the millpond. These gates would automatically close by natural force of the water current turning at ebb tide. This impounded both the tidewater and any incoming fresh-water from behind. As mentioned earlier, once the tide dropped below the level of the water inside the dam, the trapped water could be released through a sluice to fall on the wheel and set it in motion.

The most efficient operation of the mill occurred when the tide fell to a point below the level of the entire waterwheel, allowing it to "run clear." The wheel would continue to turn until either the water behind the dam fell below the level of the sluice, or more frequently, until the water level in front of the dam rose above the sluice at high tide.

"Even in coastal areas with freshwater falls nearby, tide mills offered the unique advantage of a water supply that was entirely dependable. They were typically free from the risks of drought or upstream diversions into manmade reservoirs, canals or irrigation flow of water into ditches. Compared to windmills, tide mills had the obvious advantage of not depending on the strength or direction of the wind on a given day. Indeed, too-strong winds were a hazard to the sails of windmills. They were often cheaper to locate and build than water mills because no dam was needed; however, they were usually more expensive to keep in good working order over the long term.

"Of course, tide mills had the major disadvantage that the tides, while predictable, occurred at different times of the day. Humans naturally follow the sun to determine their activities, but tide millers worked according to a tidal calendar chiefly determined by the moon. The lunar day being 24 hours and 50 minutes long placed the tide miller among the earliest categories of rotating-shift workers. At harvest and other peak times, all-night duties were common. A tide miller would need to split his "full-night's sleep" into nap periods during the twice-daily incoming tides".

The working cycle of Tide Mill

Phase 1

At high tide, when the millpond had filled, the intake sluice gate and penstock sluice gate were closed.

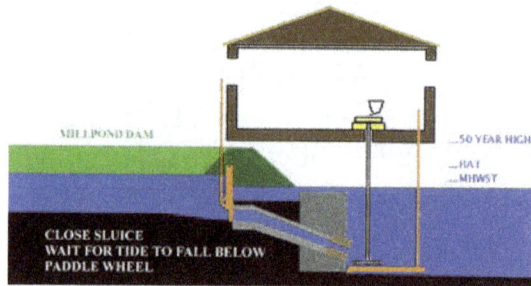

Waiting for tide to fall below wheel

Phase 2

When the water level in the wheelhouse and tailrace had fallen below the water wheel (about four hours later), the penstock sluice gate was opened, allowing the jet to commence driving the waterwheel and continue until the rising tide in the wheelhouse reached the wheel once more. This working period would, again, have been about four hours.

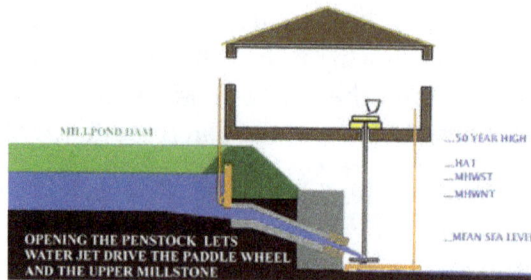

Opening the penstock to drive the wheel

Phase 3

Waiting, with the millpond intake sluice gate open, until the rising tide had filled the millpond again.

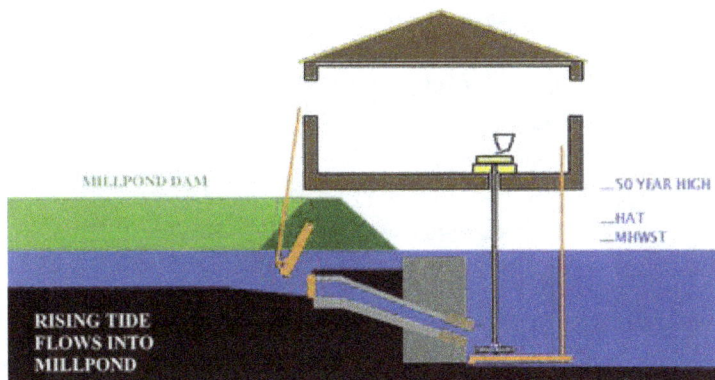

Waiting for rising tide to refill mill pond

Gristmill

A gristmill (also: grist mill, corn mill or flour mill) grinds cereal grain into flour and middlings. The term can refer to both the grinding mechanism and the building that holds it.

The early mills had horizontal paddle wheels, an arrangement which later became known as the "Norse wheel", as many were found in Scandinavia. The paddle wheel was attached to a shaft which was, in turn, attached to the centre of the millstone called the "runner stone". The turning force produced by the water on the paddles was transferred directly to the runner stone, causing it to grind against a stationary "bed", a stone of a similar size and shape. This simple arrangement required no gears, but had the disadvantage that the speed of rotation of the stone was dependent on the volume and flow of water available and was, therefore, only suitable for use in mountainous regions with fast-flowing streams. This dependence on the volume and speed of flow of the water also meant that the speed of rotation of the stone was highly variable and the optimum grinding speed could not always be maintained.

Vertical wheels were in use in the Roman Empire by the end of the first century BC, and these were described by Vitruvius. The peak of Roman technology is probably the Barbegal aqueduct and mill where water with a 19-metre fall drove sixteen water wheels, giving a grinding capacity estimated at 2.4 to 3.2 tonnes per hour. Water mills seem to have remained in use during the post-Roman period, and by 1000 AD, mills in Europe were rarely more than a few miles apart.

The old water mill at Decew Falls

In England, the Domesday survey of 1086 gives a precise count of England's water-powered flour mills: there were 5,624, or about one for every 300 inhabitants, and this was probably typical throughout western and southern Europe. From this time onward, water wheels began to be used for purposes other than grist milling. In England, the number of mills in operation followed population growth, and peaked at around 17,000 by 1300.

Limited extant examples of gristmills can be found in Europe from the High Middle Ages. An extant well-preserved waterwheel and gristmill on the Ebro River in Spain is associated with the Real Monasterio de Nuestra Senora de Rueda, built by the Cistercian monks in 1202. The Cistercians were known for their use of this technology in Western Europe in the period 1100 to 1350.

Geared gristmills were also built in the medieval Near East and North Africa, which were used for grinding grain and other seeds to produce meals. Gristmills in the Islamic world were powered by both water and wind. The first wind-powered gristmills were built in the 9th and 10th centuries in what are now Afghanistan, Pakistan and Iran.

Classical British and American mills

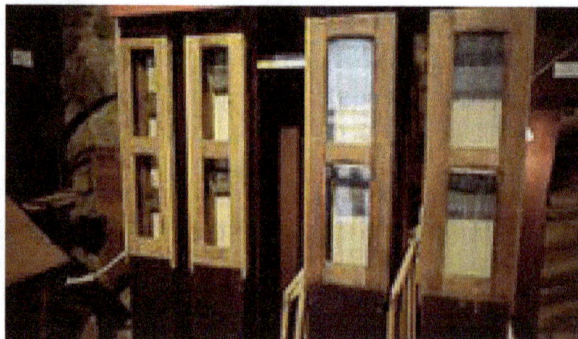

Wayside Inn Grist Mill in Massachusetts

Stretton Watermill, 17th-century built operational mill in Cheshire, England

Although the terms "gristmill" or "corn mill" can refer to any mill that grinds grain, the terms were used historically for a local mill where farmers brought their own grain and received back ground meal or flour, minus a percentage called the "miller's toll." Early mills were almost always built and supported by farming communities and the miller received the "miller's toll" in lieu of wages. Most towns and villages had their own mill so that local farmers could easily transport their grain there to be milled. These communities were dependent on their local mill as bread was a staple part of the diet.

Classical mill designs are usually water-powered, though some are powered by the wind or by livestock. In a watermill a sluice gate is opened to allow water to flow onto, or under, a water wheel to make it turn. In most watermills the water wheel was mounted vertically, i.e., edge-on, in the water, but in some cases horizontally (the tub wheel and so-called Norse wheel). Later designs incorporated horizontal steel or cast iron turbines and these were sometimes refitted into the old wheel mills.

In most wheel-driven mills, a large gear-wheel called the *pit wheel* is mounted on the same axle as the water wheel and this drives a smaller gear-wheel, the *wallower*, on a main driveshaft running vertically from the bottom to the top of the building. This system of gearing ensures that the main shaft turns faster than the water wheel, which typically rotates at around 10 rpm.

The millstones themselves turn at around 120 rpm. They are laid one on top of the other. The bottom stone, called the *bed*, is fixed to the floor, while the top stone, the *runner*, is mounted on a separate spindle, driven by the main shaft. A wheel called the *stone nut* connects the runner's spindle to the main shaft, and this can be moved out of the way to disconnect the stone and stop it turning, leaving the main shaft turning to drive other machinery. This might include driving a mechanical sieve to refine the flour, or turning a wooden drum to wind up a chain used to hoist sacks of grain to the top of the mill house. The distance between the stones can be varied to produce the grade of flour required; moving the stones closer together produces finer flour.

The grain is lifted in sacks onto the *sack floor* at the top of the mill on the hoist. The sacks are then emptied into bins, where the grain falls down through a hopper to the millstones on the *stone floor* below. The flow of grain is regulated by shaking it in a gently sloping trough (the *slipper*) from which it falls into a hole in the center of the runner stone. The milled grain (flour) is collected as it emerges through the grooves in the runner stone from the outer rim of the stones and is fed down a chute to be collected in sacks on the ground or *meal* floor. A similar process is used for grains such as wheat to make flour, and for maize to make corn meal.

In order to prevent the vibrations of the mill machinery from shaking the building apart, a gristmill will often have at least two separate foundations.

American inventor Oliver Evans revolutionized this labor-intensive process at the end of the eighteenth century when he patented and promoted a fully automated mill design.

Modern Mills

Modern mills are highly automated. Interior in Tartu Mill, that is the biggest grain milling company in the Baltic states.

The Pilgrim's Pride feed mill in Pittsburg, Texas

Modern mills typically use electricity or fossil fuels to spin heavy steel, or cast iron, serrated and flat rollers to separate the bran and germ from the endosperm. The endosperm is ground to create white flour, which may be recombined with the bran and germ to create whole grain or graham flour. The different milling techniques produce visibly different results, but can be made to produce nutritionally and functionally equivalent output. Stone-ground flour is, however, preferred by many bakers and natural food advocates because of its texture, nutty flavor, and the belief that it is nutritionally superior and has a better baking quality than steel-roller-milled flour. It is claimed that, as the stones grind relatively slowly, the wheat germ is not exposed to the sort of excessive temperatures that could cause the fat from the germ portion to oxidize and become rancid, which would destroy some of the vitamin content. Stone-milled flour has been found to be relatively high in thiamin, compared to roller-milled flour, especially when milled from hard wheat.

Gristmills only grind "clean" grains from which stalks and chaff have previously been removed, but historically some mills also housed equipment for threshing, sorting, and cleaning prior to grinding.

Modern mills are usually "merchant mills" that are either privately owned and accept money or trade for milling grains or are owned by corporations that buy un-milled grain and then own the flour produced.

Sawmill

An American sawmill, circa 1920

Early 20th-century sawmill, maintained at Jerome, Arizona

A sawmill or lumber mill is a facility where logs are cut into lumber. Before the invention of the sawmill, boards were made in various manual ways, either rived (split) and planed, hewn, or more often hand sawn by two men with a whipsaw, one above and another in a saw pit below. The earliest known mechanical mill is the Hierapolis sawmill, a Roman water-powered stone mill at Hierapolis, Asia Minor dating back to the 3rd century AD. Other water-powered mills followed and by the 11th century they were widespread in Spain and North Africa, the Middle East and Central Asia, and in the next few centuries, spread across Europe. The circular motion of the wheel was converted to a reciprocating motion at the saw blade. Generally, only the saw was powered, and the logs had to be loaded and moved by hand. An early improvement was the development of a movable carriage, also water powered, to move the log steadily through the saw blade.

By the time of the Industrial Revolution in the 18th century, the circular saw blade had been invented, and with the development of steam power in the 19th century, a much greater degree of mechanisation was possible. Scrap lumber from the mill provided a source of fuel for firing the boiler. The arrival of railroads meant that logs could be transported to mills rather than mills being built besides navigable waterways. By 1900, the largest sawmill in the world was operated by the Atlantic Lumber Company in Georgetown, South Carolina, using logs floated down the Pee Dee River from the Appalachian Mountains. In the 20th century the introduction of electricity and high technology furthered this process, and now most sawmills are massive and expensive facilities in which most aspects of the work is computerized. Besides the sawn timber, use is made of all the by-products including sawdust, bark, woodchips, and wood pellets, creating a diverse offering of forest products.

Sawmill Process

A sawmill's basic operation is much like those of hundreds of years ago; a log enters on one end and dimensional lumber exits on the other end.

- After trees are selected for harvest, the next step in logging is felling the trees, and bucking them to length.

- Branches are cut off the trunk. This is known as *limbing*.

- Logs are taken by logging truck, rail or a log drive to the sawmill.

- Logs are scaled either on the way to the mill or upon arrival at the mill.

- Debarking removes bark from the logs.

- *Decking* is the process for sorting the logs by species, size and end use (lumber, plywood, chips).

- A sawyer uses a head saw (also called head rig or primary saw) to break the log into cants (unfinished logs to be further processed) and flitches (unfinished planks).

- Depending upon the species and quality of the log, the cants will either be further broken down by a resaw or a gang edger into multiple flitches and/or boards.

- *Edging* will take the flitch and trim off all irregular edges leaving four-sided lumber.

- *Trimming* squares the ends at typical lumber lengths.

- *Drying* removes naturally occurring moisture from the lumber. This can be done with kilns or air-dried.

- *Planing* smooths the surface of the lumber leaving a uniform width and thickness.

- *Shipping* transports the finished lumber to market.

Pre-Industrial Revolution

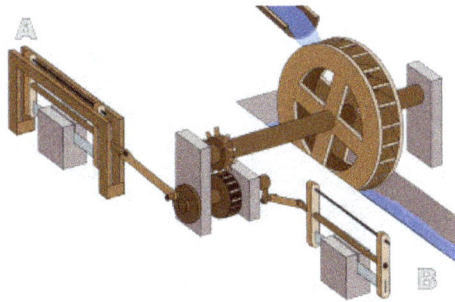

Scheme of the water-driven sawmill at Hierapolis, Asia Minor.
The 3rd-century mill incorporated a crank and connecting rod mechanism.

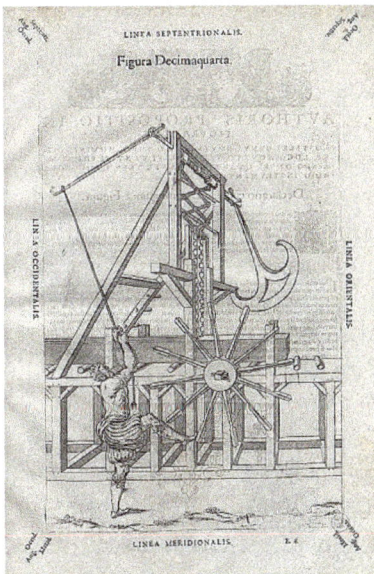

Illustration of a human-powered sawmill with a gang-saw

"De Salamander" a wind driven sawmill in Leidschendam, The Netherlands.

A sawmill in the interior of Australia

Modern reconstruction Sutter's mill in California

The Hierapolis sawmill, a water-powered stone saw mill at Hierapolis, Asia Minor (modern-day Turkey, then part of the Roman Empire), dating to the second half of the 3rd century, is the earliest known sawmill. It also incorporates a crank and connecting rod mechanism.

Water-powered stone sawmills working with cranks and connecting rods, but without gear train, are archaeologically attested for the 6th century at the Byzantine cities Gerasa (in Asia Minor) and Ephesus (in Syria).

The earliest literary reference to a working sawmill comes from a Roman poet, Ausonius, who wrote a topographical poem about the river Moselle in Germany in the late 4th century AD. At one point in the poem he describes the shrieking sound of a water-mill cutting marble. Marble sawmills also seem to be indicated by the Christian saint Gregory of Nyssa from Anatolia around 370/390 AD, demonstrating a diversified use of water-power in many parts of the Roman Empire.

By the 11th century, hydropowered sawmills were in widespread use in the medieval Islamic world, from Islamic Spain and North Africa in the west to Central Asia in the east.

Sawmills later became widespread in medieval Europe, as one was sketched by Villard de Honnecourt in c. 1250. They are claimed to have been introduced to Madeira following its discovery in c. 1420 and spread widely in Europe in the 16th century.

Prior to the invention of the sawmill, boards were rived (split) and planed, or more often sawn by two men with a whipsaw, using saddleblocks to hold the log, and a saw pit for the pitman who worked below. Sawing was slow, and required strong and hearty men. The topsawer had to be the stronger of the two because the saw was pulled in turn by each man, and the lower had the advantage of gravity. The topsawyer also had to guide the saw so that the board was of even thickness. This was often done by following a chalkline.

Early sawmills simply adapted the whipsaw to mechanical power, generally driven by a water wheel to speed up the process. The circular motion of the wheel was changed to back-and-forth motion of the saw blade by a connecting rod known as a *pitman arm*.

Generally, only the saw was powered, and the logs had to be loaded and moved by hand. An early improvement was the development of a movable carriage, also water powered, to move the log steadily through the saw blade.

A type of sawmill without a crank is known from Germany called "knock and drop" or simply "drop" -mills. In these drop sawmills, the frame carrying the saw blade is knocked upwards by cams as the shaft turns. These cams are let into the shaft on which the waterwheel sits. When the frame carrying the saw blade is in the topmost position it drops by its own weight, making a loud knocking noise, and in so doing it cuts the trunk.

A small mill such as this would be the center of many rural communities in wood-exporting regions such as the Baltic countries and Canada. The output of such mills would be quite low, perhaps only 500 boards per day. They would also generally only operate during the winter, the peak logging season.

In the United States, the sawmill was introduced soon after the colonisation of Virginia by recruiting skilled men from Hamburg. Later the metal parts were obtained from the Netherlands, where the technology was far ahead of that in England, where the sawmill remained largely unknown until the late 18th century. The arrival of a sawmill was a large and simulative step in the growth of a frontier community.

Industrial Revolution

Early mills had been taken to the forest, where a temporary shelter was built, and the logs were skidded to the nearby mill by horse or ox teams, often when there was some snow to provide lubrication. As mills grew larger, they were usually established in more permanent facilities on a river, and the logs were floated down to them by log drivers. Sawmills built on navigable rivers, lakes, or estuaries were called cargo mills because of the availability of ships transporting cargoes of logs to the sawmill and cargoes of lumber from the sawmill.

The next improvement was the use of circular saw blades, perhaps invented in England in the late 18th century, but perhaps in 17th-century Holland, the Netherlands. Soon thereafter, millers used gangsaws, which added additional blades so that a log would be reduced to boards in one quick step. Circular saw blades were extremely expensive and highly subject to damage by overheating or dirty logs. A new kind of technician arose, the sawfiler. Sawfilers were highly skilled in metalworking. Their main job was to *set* and sharpen teeth. The craft also involved learning how to *hammer* a saw, whereby a saw is deformed with a hammer and anvil to counteract the forces of heat and cutting. Modern circular saw blades have replaceable teeth, but still need to be hammered.

The introduction of steam power in the 19th century created many new possibilities for mills. Availability of railroad transportation for logs and lumber encouraged building of rail mills away from navigable water. Steam powered sawmills could be far more mechanized. Scrap lumber from the mill provided a ready fuel source for firing the boiler. Efficiency was increased, but the capital cost of a new mill increased dramatically as well.

In addition, the use of steam or gasoline-powered traction engines also allowed the entire sawmill to be mobile.

By 1900, the largest sawmill in the world was operated by the Atlantic Lumber Company in Georgetown, South Carolina, using logs floated down the Pee Dee River from as far as the edge of the Appalachian Mountains in North Carolina.

A restoration project for Sturgeon's Mill in Northern California is underway, restoring one of the last steam-powered lumber mills still using its original equipment.

Current Trends

Oregon Mill using energy efficient ponding to move logs

In the twentieth century the introduction of electricity and high technology furthered this process, and now most sawmills are massive and expensive facilities in which most aspects of the work is computerized. The cost of a new facility with 2 Mmfbm/day capacity is up to CAN$120,000,000. A modern operation will produce between 100 Mmfbm and 700 Mmfbm annually.

Small gasoline-powered sawmills run by local entrepreneurs served many communities in the early twentieth century, and specialty markets still today.

A trend is the small portable sawmill for personal or even professional use. Many different models have emerged with different designs and functions. They are especially suitable for producing limited volumes of boards, or specialty milling such as oversized timber. Portable sawmills have gained popularity for the convenience of bringing the sawmill to the logs and milling lumber in remote locations. Some remote communities that have experienced natural disasters have used portable sawmills to rebuild their communities out of the fallen trees.

Technology has changed sawmill operations significantly in recent years, emphasizing increasing profits through waste minimization and increased energy efficiency as well as improving operator safety. The once-ubiquitous rusty, steel conical sawdust burners have for the most part vanished, as the sawdust and other mill waste is now processed into particleboard and related products, or used to heat wood-drying kilns. Co-generation facilities will produce power for the operation and may also feed superfluous energy onto the grid. While the bark may be ground for landscaping barkdust, it may also be burned for heat. Sawdust may make particle board or be pressed into wood pellets for pellet stoves. The larger pieces of wood that won't make lumber are chipped into wood chips and provide a source of supply for paper mills. Wood by-products of the mills will also make oriented strand board (OSB) paneling for building construction, a cheaper alternative to plywood for paneling. Some automatic mills can process 800 small logs into bark chips, wood chips, sawdust and sorted, stacked, and bound planks, in an hour.

Hierapolis Sawmill

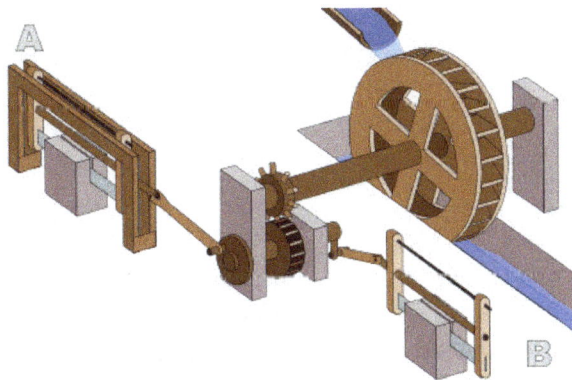

Imaginary scheme of the water-driven sawmill at Hierapolis, Asia Minor, a 3rd century mill.

The Hierapolis sawmill is believed to be a water-powered stone sawmill at Hierapolis, Asia Minor (modern-day Turkey). Dating to the second half of the 3rd century, the sawmill is considered the earliest known machine to combine a crank with a connecting rod, although neither clear ancient scripts nor engineering drawings were yet found to support this theory.

Some archaeologists believe that the watermill is evidenced by a raised relief on the sarcophagus of a certain Marcus Aurelius Ammianos, a local miller. On the pediment a waterwheel fed by a mill race is shown powering via a gear train two frame saws cutting rectangular blocks by the way of connecting rods and, through mechanical necessity, cranks. The accompanying inscription is in Greek and attributes the mechanism to Ammianos' "skills with wheels".

Other Sawmills

Further crank and connecting rod mechanisms, without gear train, are archaeologically attested for the 6th century water-powered stone sawmills at Gerasa, Jordan, and Ephesus, Turkey, both part of the Byzantine Empire at the time. A fourth sawmill possibly existed at Augusta Raurica, Switzerland, where a metal crank from the 2nd century has been excavated.

Literary references to water-powered marble saws in Trier, now Germany, can be found in Ausonius' late 4th century poem *Mosella*. About the same time, they also seem to be indicated by the Christian saint Gregory of Nyssa from Anatolia, demonstrating a diversified use of water-power in parts of the Roman Empire.

The three supposed finds push back the date of the invention of the crank and connecting rod mechanism by a full millennium. According to Tullia Ritti, Klaus Grewe and Paul Kessener:

With the crank and connecting rod system, all elements for constructing a steam engine (invented in 1712) — Hero's aeolipile (generating steam power), the cylinder and piston (in metal force pumps), non-return valves (in water pumps), gearing (in water mills and clocks) — were known in Roman times.

Blade Mill

Blade Mills, also known as Aggregate Conditioners, scour, abrade and break down water-soluble silts and deleterious clays from contaminated ores, coarse rock or sand feeds. They are most often placed ahead of wash screens to improve screening efficiencies or in front of other wash equipment to improve the Sand Equivalency (SE). Despite being very similar in appearance to Coarse Material Screw Washers, Blade Mills process material differently. For instance, all material that enters the box, must exit through the discharge opening as there is no overflow weir. Without an overflow weir, material must be discharged into another piece of process equipment and cannot be put on a conveyor belt.

Blade Mills are designed to do just that — condition. While Blade Mills are very similar in appearance to Coarse Material Screw Washers, they function much differently. They can accept both fine and coarse material, but are not designed to remove tough, plastic clays. They also usually work in conjunction with another type of processing equipment, such as a Classifying Tank or wash screen. The major difference is that any material and water that enters the Blade Mill must exit through the discharge opening located at the bottom of the box opposite the incoming feed end. There is no overflow in this type of unit.

Using a combination of paddles and flights arranged in alternating format the entire length of the shaft, they scour, abrade and break down deleterious material. The shafts can have different configurations, but mainly have alternating flights and paddles. Blade Mills sit on a slope of 0 to 5 degrees and have higher capacities compared to same size Coarse Material Screw Washers. This is primarily due to the lower machine operating slope of the Blade Mill.

Generally, the amount of water needed for a Blade Mill is an additional third of the weight of material being processed.

Again, a Blade Mill is used where the material needs to be worked and wetted before it enters other wet processing, such as a Vibrating Screen, Fine Material Screw Washer, Sand Classifying Tank or further wet processing.

References

- Oleson, John Peter (30 Jun 1984). Greek and Roman mechanical water-lifting devices: the history of a technology. Springer. p. 373. ISBN 90-277-1693-5. ASIN 9027716935.

- Paddle-wheel-technology: flowmeters.com, Retrieved 27 July 2018

- Peterson, Charles E. (1973). "Sawdust Trail: Annals of Sawmilling and the Lumber Trade from Virginia to Hawaii via Maine, Barbados, Sault Ste. Marie, Manchac and Seattle to the Year 1860". Bulletin of the Association for Preservation Technology. 5 (2): 84–153. doi:10.2307/1493399. JSTOR 1493399.

- How-do-paddle-wheel-flowmeters-work: davis.com, Retrieved 16 April 2018

- Denny, Mark (4 May 2007). "Waterwheels and Windmills". Ingenium: five machines that changed the world. The Johns Hopkins University Press. pp. 36–38. ISBN 0-8018-8586-8. Retrieved 15 December 2009.

- Paddle-wheel-technology: flowmeters.com, Retrieved 24 June 2018

- Adam Lucas (2006), Wind, Water, Work: Ancient and Medieval Milling Technology, p. 65, Brill Publishers, ISBN 90-04-14649-0

- Blade-mills, products: mclanahan.com, Retrieved 22 May 2018

- "ARTFL Project: Webster Dictionary, 1913". The University of Chicago - Department of Romance Languages and Literature. Archived from the original on 2007-03-13. Retrieved 2006-09-28.

5

Technologies used in Hydropower

A number of technologies and techniques are used in generating hydropower. This chapter discusses and analyzes energy generation through dams and the concepts of ocean thermal energy conversion, tidal barrage, tidal current, etc. in detail.

Dams

A dam is a barrier across flowing water that obstructs, directs, or slows the flow, often creating a reservoir, lake, or impoundment. In Australian and South African English, the word dam may refer to the reservoir as well as the barrier. Most dams have a section called a *spillway or weir* that allows water to flow out, either intermittently or continuously.

A dam may serve one or more purposes, such as to provide water for neighboring towns, farms, and industries, to produce hydroelectric power, to improve navigation, to control flooding, and to maintain wildlife habitats. On the downside, a dam may adversely affect the area's ecosystem and destabilize geological formations. The construction of a large dam may force the relocation of many local inhabitants, and structural failure of a dam can have catastrophic effects. The building and maintenance of a dam needs to take these factors into consideration.

Types of Dams

Dams can be formed by human agency or natural causes, including the intervention of wildlife such as beavers. Man-made dams are typically classified according to their size (height), intended purpose, or structure.

Classification by Size

International standards define *large dams* as those higher than 15 meters and *major dams* as those over 150 meters in height.

Classification by Purpose

A dam may be constructed for one or more purposes, such as:

- To provide water for irrigation;

- To provide water supply for a town or city;

- To improve navigation;

- To create a reservoir of water for industrial uses;

- To generate hydroelectric power;

- To create recreational areas;

- To maintain a habitat for fish and wildlife;

- To control floods;

- To contain effluents from sites such as mines or factories.

Few dams serve all these purposes, but some serve more than one purpose.

A *saddle dam* is an auxiliary dam constructed to confine the reservoir created by a primary dam either to permit a higher water elevation and storage or to limit the extent of a reservoir for increased efficiency. An auxiliary dam is constructed in a low spot or *saddle* through which the reservoir would otherwise escape. On occasion, a reservoir is contained by a similar structure called a dike to prevent inundation of nearby land. Dikes are commonly used for *reclamation* of arable land from a shallow lake. This is similar to a levee, which is a wall or embankment built along a river or stream to protect adjacent land from flooding.

An *overflow dam* is designed for water to flow over its top. A weir is a type of small overflow dam that can be used for flow measurement.

A *check dam* is a small dam designed to reduce flow velocity and control soil erosion. Conversely, a *wing dam* is a structure that only partly restricts a waterway, creating a faster channel that resists the accumulation of sediment.

A *dry dam* is a dam designed to control flooding. It normally holds back no water and allows the channel to flow freely, except during periods of intense flow that would otherwise cause flooding downstream.

A *diversionary dam* is a structure designed to divert all or a portion of the flow of a river from its natural course.

Classification by Structure

Based on structure and material used, dams are classified as timber dams, embankment dams or masonry dams, with several subtypes.

Masonry Dams

Arch Dams

Hoover Dam, a concrete gravity-arch dam in the Black Canyon of the Colorado River.

In the arch dam, stability is obtained by a combination of arch and gravity action. If the upstream face is vertical the entire weight of the dam must be carried to the foundation by gravity, while the distribution of the normal hydrostatic pressure between vertical cantilever and arch action will depend upon the stiffness of the dam in a vertical and horizontal direction. When the upstream face is sloped the distribution is more complicated. The normal component of the weight of the arch ring may be taken by the arch action, while the normal hydrostatic pressure will be distributed as described above. For this type of dam, firm reliable supports at the abutments (either buttress or canyon side wall) are more important. The most desirable place for an arch dam is a narrow canyon with steep side walls composed of sound rock. The safety of an arch dam is dependent on the strength of the side wall abutments, hence not only should the arch be well seated on the side walls but also the character of the rock should be carefully inspected.

Two types of single-arch dams are in use, namely the constant-angle and the constant-radius dam. The constant-radius type employs the same face radius at all elevations of the dam, which means that as the channel grows narrower towards the bottom of the dam the central angle subtended by the face of the dam becomes smaller. Jones Falls Dam, in Canada, is a constant radius dam. In a constant-angle dam, also known as a variable radius dam, this subtended angle is kept a constant and the variation in distance between the abutments at various levels is taken care of by varying the radii. Constant-radius dams are much less common than constant-angle dams. Parker Dam is a constant-angle arch dam.

A similar type is the double-curvature or thin-shell dam. Wildhorse Dam near Mountain City, Nevada in the United States is an example of the type. This method of construction minimizes the amount of concrete necessary for construction but transmits large loads to the foundation and abutments. The appearance is similar to a single-arch dam but with a distinct vertical curvature to it as well lending it the vague appearance of a concave lens as viewed from downstream.

The multiple-arch dam consists of a number of single-arch dams with concrete buttresses as the supporting abutments. The multiple-arch dam does not require as many buttresses as the hollow gravity type, but requires good rock foundation because the buttress loads are heavy.

Gravity Dams

The Gilboa Dam in the Catskill Mountains of New York State
is an example of a "solid" gravity dam.

In a gravity dam, stability is secured by making it of such a size and shape that it will resist overturning, sliding and crushing at the toe. The dam will not overturn provided that the moment around the turning point, caused by the water pressure is smaller than the moment caused by the weight of the dam. This is the case if the resultant force of water pressure and weight falls within the base of the dam. However, in order to prevent tensile stress at the upstream face and excessive compressive stress at the downstream face, the dam cross section is usually designed so that the resultant falls within the middle at all elevations of the cross section (the core). For this type of dam, impervious foundations with high *bearing* strength are essential.

When situated on a suitable site, a gravity dam inspires more confidence in the layman than any other type; it has mass that lends an atmosphere of permanence, stability, and safety. When built on a carefully studied foundation with stresses calculated from completely evaluated loads, the gravity dam probably represents the best developed example of the art of dam building. This is significant because the fear of flood is a strong motivator in many regions, and has resulted in gravity dams being built in some instances where an arch dam would have been more economical.

Gravity dams are classified as "solid" or "hollow." The solid form is the more widely used of the two, though the hollow dam is frequently more economical to construct.

Gravity dams can also be classified as "overflow" (spillway) and "non-overflow." Grand Coulee Dam is a solid gravity dam and Itaipu Dam is a hollow gravity dam.

Embankment Dams

The San Luis Dam near Los Banos, California is an embankment dam

Embankment dams are made from compacted earth, and have two main types, rock-fill and earth-fill dams. Embankment dams rely on their weight to hold back the force of water, like the gravity dams made from concrete.

Rock-fill Dams

Rock-fill dams are embankments of compacted free-draining granular earth with an impervious zone. The earth utilized often contains a large percentage of large particles hence the term *rock-fill*. The impervious zone may be on the upstream face and made of masonry, concrete, plastic membrane, steel sheet piles, timber or other material. The impervious zone may also be within the embankment in which case it is referred to as a *core*. In the instances where clay is utilized as the impervious material the dam is referred to as a *composite* dam. To prevent internal erosion of clay into the rock fill due to seepage forces, the core is separated using a filter. Filters are specifically graded soil designed to prevent the migration of fine grain soil particles. When suitable material is at hand, transportation is minimized leading to cost savings during construction. Rock-fill dams are resistant to damage from earthquakes. However, inadequate quality control during construction can lead to poor compaction and sand in the embankment which can lead to liquefaction of the rock-fill during an earthquake. Liquefaction potential can be reduced by keeping susceptible material from being saturated, and by providing adequate compaction during construction. An example of a rock-fill dam is New Melones Dam in California.

Earth-fill Dams

Earth-fill dams, also called earthen, rolled-earth or simply earth dams, are constructed as a simple embankment of well compacted earth. A *homogeneous* rolled-earth dam is entirely constructed of one type of material but may contain a drain layer to collect *seep* water. A *zoned-earth* dam has distinct parts or *zones* of dissimilar material, typically a locally plentiful *shell* with a watertight clay core. Modern zoned-earth embankments employ filter and drain zones to collect and remove seep water and preserve the integrity of the downstream shell zone. An outdated method of zoned earth dam construction utilized a hydraulic fill to produce a watertight core. *Rolled-earth* dams may also employ a watertight facing or core in the manner of a rock-fill dam. An interesting type of temporary earth dam occasionally used in high latitudes is the *frozen-core* dam, in which a coolant is circulated through pipes inside the dam to maintain a watertight region of permafrost within it.

Because earthen dams can be constructed from materials found on-site or nearby, they can be very cost-effective in regions where the cost of producing or bringing in concrete would be prohibitive.

Asphalt-concrete Core

A third type of embankment dam is built with asphalt concrete core. The majority of such dams are built with rock and or gravel as main fill material. Almost 100 dams of this design have now been built world- wide since the first dam was completed in 1962. All dams built have an excellent performance record. This type of asphalt is a viscoelastic, plastic material that can adjust to the movements and deformations imposed on the embankment as a whole and to settlements in the foundation. The flexible properties of the asphalt make such dams especially suited in earthquake regions.

Cofferdams

A cofferdam during the construction of locks at the Montgomery Point Lock and Dam

A cofferdam is a (usually temporary) barrier constructed to exclude water from an area that is normally submerged. Made commonly of wood, concrete or steel sheet piling, cofferdams are used to allow construction on the foundation of permanent dams, bridges, and similar structures. When the project is completed, the cofferdam may be demolished or removed. See also causeway and retaining wall. Common uses for cofferdams include construction and repair of off shore oil platforms. In such cases the cofferdam is fabricated from sheet steel and welded into place under water. Air is pumped into the space, displacing the water allowing a dry work environment below the surface. Upon completion the cofferdam is usually deconstructed unless the area requires continuous maintenance.

Timber Dams

A timber crib dam in Michigan

Timber dams were widely used in the early part of the industrial revolution and in frontier areas due to ease and speed of construction. Rarely built in modern times by humans due to relatively short lifespan and limited height to which they can be built, timber dams must be kept constantly wet in order to maintain their water retention properties and limit deterioration by rot, similar to a barrel. The locations where timber dams are most economical to build are those where timber is plentiful, cement is costly or difficult to transport, and either a low head diversion dam is required or longevity is not an issue. Timber dams were once numerous, especially in the North American west, but most have failed, been hidden under earth embankments or been replaced with entirely new structures. Two common variations of timber dams were the *crib* and the *plank*.

Timber crib dams were erected of heavy timbers or dressed logs in the manner of a log house and the interior filled with earth or rubble. The heavy crib structure supported the dam's face and the weight of the water.

Timber plank dams were more elegant structures that employed a variety of construction methods utilizing heavy timbers to support a water retaining arrangement of planks.

Very few timber dams are still in use. Timber, in the form of sticks, branches and withes, is the basic material used by beavers, often with the addition of mud or stones.

Steel Dams

Red Ridge steel dam

A steel dam is a type of dam briefly experimented with in around the turn of the nine-teenth-twentieth century which uses steel plating (at an angle) and load bearing beams as the structure. Intended as permanent structures, steel dams were an (arguably failed) experiment to determine if a construction technique could be devised that was cheaper than masonry, concrete or earthworks, but sturdier than timber crib dams.

Beaver Dams

Beavers create dams primarily out of mud and sticks to flood a particular habitable area. By flooding a parcel of land, beavers can navigate below or near the surface and remain relatively well hidden or protected from predators. The flooded region also al-lows beavers access to food, especially during the winter.

Construction Elements

Power Generation Plant

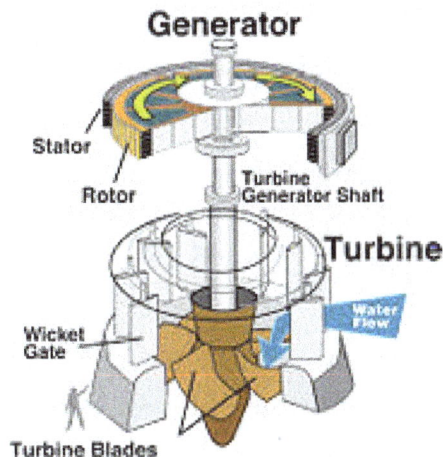

Hydraulic turbine and electrical generator.

As of 2005, hydroelectric power, mostly from dams, supplies some 19 percent of the world's electricity, and over 63 percent of renewable energy. Much of this is generated by large dams, although China uses small-scale hydro generation in many locations and is responsible for about 50 percent of world use of this type of power.

Most hydroelectric power comes from the potential energy of dammed water driving a water turbine and generator; to boost the power generation capabilities of a dam, the water may be run through a large pipe called a penstock before the turbine. A variant on this simple model uses pumped storage hydroelectricity to produce electricity to match periods of high and low demand, by moving water between reservoirs at different elevations. At times of low electrical demand, excess generation capacity is used to pump water into the higher reservoir. When there is higher demand, water is released back into the lower reservoir through a turbine.

Spillways

Spillway on Llyn Brianne dam, Wales soon after first fill

A *spillway* is a section of a dam designed to pass water from the upstream side of a dam to the downstream side. Many spillways have floodgates designed to control the flow through the spillway. Types of spillway include: A *service spillway* or *primary spillway* passes normal flow. An *auxiliary spillway* releases flow in excess of the capacity of the service spillway. An *emergency spillway* is designed for extreme conditions, such as a serious malfunction of the service spillway. A *fuse plug spillway* is a low embankment designed to be over topped and washed away in the event of a large flood.

The spillway can be gradually eroded by water flow, including cavitation or turbulence of the water flowing over the spillway, leading to its failure. It was the inadequate design of the spillway which led to the 1889 over-topping of the South Fork Dam in Johnstown, Pennsylvania, resulting in the infamous Johnstown Flood.

Erosion rates are often monitored, and the risk is ordinarily minimized, by shaping the downstream face of the spillway into a curve that minimizes turbulent flow, such as an ogee curve.

Dam Creation

Common Purposes

Function	Example
Power generation	Hydroelectric power is a major source of electricity in the world. many countries have rivers with adequate water flow, that can be dammed for power generation purposes. For example, the Itaipu on the Paraná River in South America generates 14 GW and supplied 93 percent of the energy consumed by Paraguay and 20 percent of that consumed by Brazil as of 2005.
Stabilize water flow / irrigation	Dams are often used to control and stabilize water flow, often for agricultural purposes and irrigation. Others such as the Berg Strait dam can help to stabilize or restore the water levels of inland lakes and seas, in this case the Aral Sea.
Flood prevention	Dams such as the Blackwater dam of Webster, New Hampshire and the Delta Works are created with flood control in mind.
Land reclamation	Dams (often called dikes or levees in this context) are used to prevent ingress of water to an area that would otherwise be submerged, allowing its reclamation for human use.
Water diversion	A diversion dam diverts all or part of a river flow from its natural course into an artificial course or canal. The redirected flow may be used for irrigation, passed through hydroelectric generators, channeled into a different river, or dammed to form a reservoir.

Siting

One of the best places for building a dam is a narrow part of a deep river valley; the valley sides can then act as natural walls. The primary function of the dam's structure is to fill the gap in the natural reservoir line left by the stream channel. The sites are usually those where the gap becomes a minimum for the required storage capacity. The most economical arrangement is often a composite structure such as a masonry dam flanked by earth embankments. The current use of the land to be flooded should be dispensable.

Significant other engineering and engineering geology considerations when building a dam include:

- Permeability of the surrounding rock or soil

- Earthquake faults

- Landslides and slope stability

- Peak flood flows

- Reservoir silting

- Environmental impacts on river fisheries, forests and wildlife

- Impacts on human habitations

- Compensation for land being flooded as well as population resettlement

- Removal of toxic materials and buildings from the proposed reservoir area

Impact Assessment

The impact of a dam is assessed by several criteria:

- The benefits to human society arising from the dam, such as for agriculture, water supply, damage prevention, and electric power;

- The harm or benefits to nature and wildlife, especially fish and rare species;

- The effect on the area's geology, whether changes in water flow and levels will increase or decrease geological stability;

- The disruption of human lives, such as by relocation of peoples and the loss of archaeological and cultural artifacts underwater.

Economics

Construction of a hydroelectric plant requires a long lead-time for site studies, hydrological studies, and impact assessment, and is large scale projects by comparison to traditional power generation based upon fossil fuels. The number of sites that can be economically developed for hydroelectric production is limited; new sites tend to be far from population centers and usually require extensive power transmission lines. Hydroelectric generation can be vulnerable to major changes in the climate, including variation of rainfall, ground and surface water levels, and glacial melt, causing additional expenditure for the extra capacity to ensure sufficient power is available in low water years.

Once completed, a well-designed, well-maintained hydroelectric power plant is a comparatively cheap and reliable source of electricity. It is a renewable energy source that can be readily regulated to store water as needed and generate high power levels on demand.

Impacts of Dams

Environmental Impact

Dams affect many ecological aspects of a river. For example, a dam slows a river and affects the ecological pattern established by the river through its rate of flow. Also, rivers tend to have fairly homogeneous temperatures, but reservoirs have layered temperatures: warm on the top and cold on the bottom. In addition, because it is water from the colder (lower) layer of the reservoir that is often released downstream, it may have a different dissolved oxygen content than regular river water. Organisms depending upon a regular cycle of temperatures may be unable to adapt; the balance of other fauna (especially plant life and microscopic fauna) may be affected by the change of oxygen content.

Older dams often lack a fish ladder, preventing many fish from moving upstream to their natural breeding grounds. This leads to failure of breeding cycles and blocks migration paths. Even with the presence of a fish ladder, there may be a reduction in the number of fish reaching their upstream spawning grounds. In some areas, young fish ("smolt") are transported downstream by barge during parts of the year. Researchers are actively working on turbine and power-plant designs that could have lower impacts on aquatic life.

A large dam can cause the loss of entire ecospheres, including endangered and undiscovered species in the area, and replacement of the original environment by a new inland lake.

Water exiting a turbine usually contains very little suspended sediment, which can lead to scouring of river beds and loss of riverbanks. For example, the daily cyclic flow variation caused by the Glen Canyon Dam contributed to sand bar erosion.

Depending upon the circumstances, a dam may either increase or decrease the net production of greenhouse gases. An increase can occur if the reservoir created by the dam itself acts as a source of substantial amounts of potent greenhouse gases (such as methane and carbon dioxide), by the decay of plant material in flooded areas in an anaerobic environment. According to a report by the World Commission on Dams, when a relatively large reservoir is built with no prior clearing of forest in the flooded area, greenhouse gas emissions from the reservoir could be higher than those of a conventional, oil-fired power plant. On the other hand, a decrease of greenhouse gas emissions can occur if the dam is used in place of traditional power generation, because electricity produced from hydroelectric generation does not give rise to any flue gas emissions from fossil fuel combustion (including sulfur dioxide, nitric oxide, carbon monoxide, dust, and mercury from coal).

Social Impacts

The impacts of a dam on human society are also significant. For example, the Three

Gorges Dam on the Yangtze River in China, is more than five times the size of the Hoover Dam (USA) and will create a reservoir 600 km long, to be used for hydro-power generation. Its construction led to the loss of over a million people's homes and their mass relocation, the loss of many valuable archaeological and cultural sites, as well as significant ecological changes.

Dam Failure

The reservoir emptying through the failed Teton Dam

International special sign for works and installations containing dangerous forces

Dam failures are generally catastrophic if the structure is breached or significantly damaged. Routine monitoring of seepage from drains in and around larger dams is necessary to anticipate any problems and permit remedial action before structural failures occur. Most dams incorporate mechanisms to permit the reservoir level to be lowered or drained in the event of such problems. Another solution is rock grouting, that is, pressure-pumping portland cement slurry into weak, fractured rock.

During an armed conflict, a dam is considered an "installation containing dangerous forces," because destruction of the dam could have a massive impact on the civilian population and environment. As such, it is protected by the rules of International Humanitarian Law (IHL) and shall not be made the object of attack if that could result in severe losses among the civilian population. To facilitate the identification of a dam, a protective sign is displayed, consisting of three bright orange circles aligned along a single axis, as defined by the rules of IHL.

The main causes of dam failure include spillway design error (South Fork Dam), geological instability caused by changes to water levels during filling or poor surveying (Vajont Dam, Malpasset), poor maintenance, especially of outlet pipes (Lawn Lake Dam, Val di Stava Dam Collapse), extreme rainfall (Shakidor Dam), and human, computer, or design error (Buffalo Creek Flood, Dale Dike Reservoir, Taum Sauk pumped storage plant).

Prior to the above IHL ruling, a notable case of deliberate dam failure was the British Royal Air Force Dambusters raid on Germanyduring World War II (codenamed *"Operation Chastise"*). In that raid, three German dams were selected to be breached to have an impact on German infrastructure and manufacturing and power capabilities deriving from the Ruhr and Eder rivers. This raid later became the basis for several films.

Wave Energy Converter

Vast and reliable, wave power has long been considered as one of the most promising renewable energy sources. Wave Energy Converters (WECs) convert wave power into electricity. Although attempts to utilize this resource date back to at least 1890, wave power is currently not widely employed. The plethora of innovational ideas for wave power conversion have been invented in the last three decades, resulting in thousands of patents over recent years. At present, a number of different wave energy concepts are being investigated by companies and academic research groups around the world. Although many working designs have been developed and tested through modelling and wave tank-tests, only a few concepts have progressed to sea testing. Rapidly decreasing costs however, should enable wave plants to compete favorably with conventional power plants in the near future.

Classifications

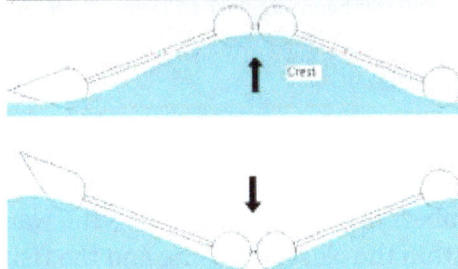

Figure: DEXA concept

Wave Activated Bodies

Wave activated bodies (WABs) are devices with moving elements that are directly activated by the cyclic oscillation of the waves. Power is extracted by converting the kinetic energy of these displacing parts into electric current. One example of such a WAB, is made by a single floater connected to a linear magnetic generator fixed to the seafloor. In other cases, only parts of the body are fully immersed and dragged by the orbital movements of the water. In order to maximally exploit this resource, the moving compounds need to be small in comparison to the wavelength and preferably they are placed half a wavelength apart. For these reasons, wave activated bodies are usually very compact and light. The main disadvantage of this type of wave energy converters is the high cost of the power generator needed to convert the irregular oscillatory flux into electricity.

The "DEXA", developed and patented by DEXA Wave Energy, is an illustrative example of a WAB. The device consists of two hinged catamarans that pivot relative to the other. The resulting oscillatory flux at the hinge, is harnessed by means of a water-based low pressure power transmission that restrains angular oscillations. Flux generation is optimized by placing the floaters of each catamaran half a wavelength apart. A scaled prototype placed in the Danish part of the North Sea should generate 160 kW. Full-scale models are thought to be able to generate up to 250 kW.

Figure : Different concepts of oscillating water columns

Oscillating Water Columns

The functioning of the oscillating water columns (OWCs) is very similar to that of a wind turbine, being based on the principle of wave induced air pressurization. The device is set upon a closed air chamber, which is placed above the water. The passage of waves changes the water level within the closed housing and the rising and falling water level increases and decreases the air pressure within the housing introducing a bidirectional air flow. By placing a turbine on top of this chamber air will pass in and out of it with the changing air pressure levels. There are two options to separate the bi-directional flow: a Wells turbine to create suction or alternatively, pressure generating valves. OWC devices can be moored offshore or be placed on the shoreline where waves break.

An example of an offshore OWC is the "Sperboy", developed and patented by Embley Energy LTD . It is circular in plane and therefore invariant to wave direction. Its size varies according to the target sea conditions at the deployment site but maximum dimensions are set at 30m diameter, 50m height and 35m draft. Up to 450 kW mean annual output can be obtained from this concept. An inshore example is the resonant wave energy converter "REWEC-3", created by the Università degli Studi "Mediterranea" di Reggio Calabria. It operates much like conventional concrete caisson breakwaters but here, each caisson is fitted with a Wells turbine. Efficiency of these devices is generally considered to be high.

Figure: Different overtopping devices

Overtopping Devices

Another type of Wave energy converter is the overtopping device, which works much like a hydroelectric dam. The "Wave Dragon" created by Wave Dragon ApS is an example of an offshore overtopping device. Its floating arms focus waves onto a slope from which the wave overtops into a reservoir. The resulting difference in water elevation between the reservoir and the mean sea level then drives low-head hydro turbines. Proposed optimal size design of 260m width and 150m length will produce 4 MW. In wave climates above 33 kW/m, this technology is expected to be economically competitive with offshore wind power in the near future. After a combined cost saving and power efficiency increase, the power price will eventually be in line with costs of fossil fuel generation.

Near shore, OVTs can be installed in front of or as part of caisson breakwaters. The Norwegian company WAV Energy is developing an integrated multilevel overtopping device named the "Sea Wave Slot-Cone Generator (SSG)". The SSG has the advantage of harvesting wave energy in several reservoirs placed above each other, resulting in high hydraulic efficiency. The reservoir capacity smooth out the irregularity of incoming waves, providing a regular electricity output to the grid. Additionally, with the turbine shaft and the gates controlling the water flow, SSG is built as a robust concrete structure with few moving parts in the mechanical system. This most likely makes it a low maintenance, durable system. Other SSG designs can be deployed onshore or offshore.

Figure: The FO3 point absorber (above) and the Wave Star attenuator

Point Absorbers and Attenuators

Point absorber are buoy-type WECs that harvest incoming wave-energy from all directions. They're placed offshore at or near the ocean surface. A vertically submerged floater absorbs wave energy which is converted by a piston or linear generator into electricity. One such a point absorber WEC is the FO3 concept developed by Norwegian entrepreneur Fred Olsen. It consists of several (12 or 21) heaving floaters attached to a 36 by 36 meter rig (Figure above). By means of a hydraulic system, the vertical motion is converted into a rotational movement that drives the hydraulic motor. This motor in turn powers the generator that can produce up to 2,52 MW.

Comparable, the attenuator type WEC "Wave Star", developed by Wave Star ApS, has a number of floaters on movable arms (Figure above). The energy of the motion of the arms is again captured in a common hydraulic line and converted into electric current. Most noticeably, being able to raise the entire installation along its pillars, this system has a high endurance for rough storm conditions. So far, this method has not been de-ployed at full scale. A 1:2 scaled installation has been built at Hanstholm which turns out 600 kW. However, production is thought to be scale-able up to 6 MW. A major ben-efit of these types of exploitation is the minimal contact with water, placing any delicate machinery and electrics out of reach of any corrosion or physical forcing of the waves.

Types of WECs

The WECs can be categorized into three types based on their size and direction of elon-gation: attenuators, point absorbers and terminators. Figure shows a schematic of these converter types. The attenuators are elongated structures with dimensions larger than the wavelength of the waves, and are oriented parallel to the wave propagation direc-tion. Each attenuator consists of a chain of cylindrical components that are connected by hydraulic pump joints so that they can conform to the local shape of the oscillatory wave. The point absorbers have dimensions much smaller than the wavelength of the waves. Unlike the attenuators, their small structure allows them to absorb wave energy from all directions. The terminators are similar to the attenuators with one main differ-ence: they are oriented perpendicular to the wave propagation direction.

Form of energy	Estimaged global resource [TWh/yr]
Waves	8,000 - 80,000
Tides	300+
Currents	800+
Thermal gradients	10,000
Salinity gradients	2,000

Table: An estimated global resource for various forms of ocean energy.

Pressure Differential Principle

As shown in figure below, the WECs that operate with variations in the air pressure inside a chamber can either be submerged, as in Archimedes effect converters, or semi-submerged, as in oscillating wave columns, under the sea water. The Archimedes effect converters (ex. Archimedes Wave Swing) are air-filled chambers in the form of point absorbers with movable upper cylinders that are moored to the sea bed. The vari-ations in pressure exerted on the upper cylinder during the crests and troughs move the cylinder down and up, and this mechanical movement is turned into electricity by a linear electric generator.

The oscillating wave columns are one of the first developments of WECs, with their floating version first developed by Yoshio Masuda in the 1960s and 1970s. The oscillat-ing wave columns (ex. Limpet) are air chambers with the bottom open to the sea water below the water free surface. As the water level inside the chamber rises and falls due to the oscillatory movement of waves, the air inside is compressed and pushed out of the chamber or expanded and sucked into the chamber through a turbine. The Wells turbine, a bidirectional turbine, is utilized so that both inflow and outflow of the air turn the turbine in the same direction.

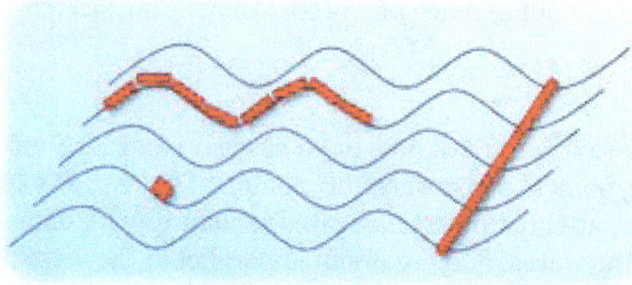

Figure: A schematic of three WEC types: an attenuator (top left),
a point absorber (bottom left) and a terminator (right).

Mechanical Flexing and Bobbing Principle

When floating structures conform to their local sea water altitudes, they move relative to each other in the case of attenuators, or with respect to the hinge point in the case of point absorbers. The wave energy can be harnessed by turning these wave-induced mechanical movements into electricity. In attenuators (ex. Pelamis), hydraulic pumps at the joints between the cylindrical components are designed to resist mechanical flexing. When the cylinders conform to the incoming wave shape as shown in figure, this mechanical flexing compresses these hydraulic pumps and pumps oil into a high-pressure tank, which then generates electricity via a hydraulic power take-off. The floating point absorbers (ex. PowerBuoy) also operate with a similar mechanism, but in this case it is the bobbing of the floating structures with waves that pumps a fluid, usually oil or sea water, and generates electricity via a hydraulic power take-off or via a linear electric generator.

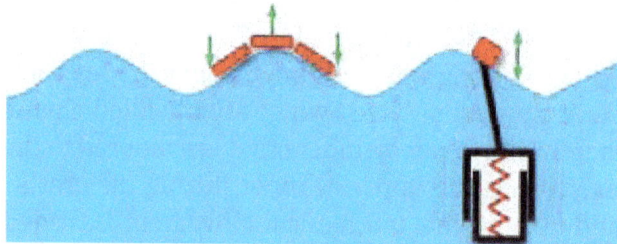

Figure: A schematic of WECs operating with the
mechanical flexing and bobbing principle.

Overtopping Principle

The wave energy can also be captured in the form of potential energy. As shown in Fig. 4, the WECs under this category are designed to capture sea water that enters a tapered channel into a reservoir raised above the sea level. The accumulated sea water in the reservoir is then controllably released back to the ocean through a hydraulic turbine to generate electricity. These WECs can either be stationed onshore (ex. Tapchan) or offshore (ex. Wave Dragon).

Hydraulic Flapping Principle

When moored terminator structures oriented perpendicular to the wave propagation direction (ex. Aquamarine Power Oyster) are hit by waves, the wave energy is absorbed upon impact and deflects the structures. As waves come in and out, this deflection generates a flapping motion as shown in figure below. Similar to the mechanical flexing and bobbing principle, the flapping movement pumps a high-pressure fluid to generate electricity via a hydraulic power take-off.

Figure: A schematic of WECs operating with the overtopping (right) and hydraulic flapping (left) principles.

Challenges

Designs are quite different from WEC to WEC, mainly due to differences in energy harvesting and subsequent conversion (Power Take-Off). Nevertheless, each design faces the same challenges. They should be optimized to effectively extract wave energy under most wave conditions while used materials are to withstand the classical problems of marine technologies i.e. corrosion, fatigue, biofouling, impact loading and fractures. The classical protection measure against fouling and corrosion of steel structures is regular maintenance and repainting. But this is time-consuming and costly. In addition, the use of antifouling paints may be detrimental to the marine environment (e.g. Tributyltin paints). Fullscale devices out of concrete (e.g. Dexa, WaveStar, SSG, RE-WEC3, Wave Dragon) may provide a valuable alternative since concrete is long-lived if properly mixed.

Crucial for any design is the mooring which ensures a maintained position under both normal operating loads as well as extreme storm load conditions. It shouldn't exert excess tension loads on the electrical transmission cables and ensure the suitable safety distances between devices in multiple installations. Most commonly, a free hanging catenary configuration is used for mooring but multi-catenary systems and flexible risers are not infrequent as well. Every configuration should be sufficiently compliant to accommodate tidal variations and environmental loading while remaining sufficiently stiff to allow berthing for inspections and maintenance. Finally, the system should be capable of lasting 30 years or more.

Wave Energy Converters as a Coastal Defense Technique

Wave Energy Converters are generally not very sensitive to sea level rise since this is expected to be small over the lifetime of most designs. The design of coastal defending WECs should however be ideally optimized to reflect and/or absorb a significant part of incident wave energy under all wave conditions, especially when presented with rough conditions. Unfortunately, the geometry of the layout which maximizes wave attenuation is yet to be determined. Existing deeply tested WECs such as the Wave Dragon and the DEXA devices are being investigated as coastal protection measures. The DEXA devices are small and therefore might be cost effective. Placement in shallow waters may reduce the transmitted energy in a differential way which could alter the coastal morphology. The Wave Dragon was chosen for its large energy absorption and reflection capacities. According to numerical simulations, the wave climate behind a single Wave Dragon and an array has significantly reduced wave heights . However, the model was not based on validated absorption and reflection performances. More detailed knowledge on the performance of a single device needs to be generated before a reliable model can be created that, ultimately, will contribute to optimization of the array lay-out.

Ocean Thermal Energy Conversion

Ocean thermal energy conversion (OTEC) is the form of energy conversion that makes use of the temperature differential between the warm surface waters of the oceans, heated by solar radiation, and the deeper cold waters to generate power in a conventional heat engine. The difference in temperature between the surface and the lower water layer can be as large as 50° C (90° F) over vertical distances of as little as 90 metres (about 300 feet) in some ocean areas. To be economically practical, the temperature differential should be at least 20° C (36° F) in the first 1,000 metres (about 3,300 feet) below the surface. In the first decade of the 21st century, the technology was still considered to be experimental, and thus far no commercial OTEC plants have been constructed.

The OTEC concept was first proposed in the early 1880s by the French engineer Jacques-Arsène d'Arsonval. His idea called for a closed-cycle system, a design that has been adapted for most present-day OTEC pilot plants. Such a system employs a secondary working fluid (a refrigerant) such as ammonia. Heat transferred from the warm surface ocean water causes the working fluid to vaporize through a heat exchanger. The vapour then expands under moderate pressures, turning a turbine connected to a generator and thereby producing electricity. Cold seawater pumped up from the ocean depths to a second heat exchanger provides a surface cool enough to cause the vapour to condense. The working fluid remains within the closed system, vaporizing and reli-quefying continuously.

Some researchers have centred their attention on an open-cycle OTEC system that employs water vapour as the working fluid and dispenses with the use of a refrigerant. In this kind of system, warm surface seawater is partially vaporized as it is injected into a near vacuum. The resultant steam is expanded through a low-pressure steam turbo generator to produce electric power. Cold seawater is used to condense the steam, and a vacuum pump maintains the proper system pressure. Hybrid systems, which combine elements of closed-cycle and open-cycle systems, also exist. In these systems, steam produced by warm water passing through a vacuum chamber is used to vaporize a secondary working fluid that drives a turbine.

The prospects for commercial application of OTEC technology seem bright, particularly on islands and in developing countries in the tropical regions where conditions are most favourable for OTEC plant operation. It has been estimated that the tropical ocean waters absorb solar radiation equivalent in heat content to that of about 250 billion barrels of oil each day. Removal of this much heat from the ocean would not significantly alter its temperature, but it would permit the generation of tens of millions of megawatts of electricity on a continuous basis.

Beyond the production of clean power, the OTEC process also provides several useful by-products. The delivery of cool water to the surface has been used in air-conditioning systems and in chilled-soil agriculture (which allows for the cultivation of temperate-zone plants in tropical environments). Open-cycle and hybrid processes have been used in seawater desalination, and OTEC infrastructure allows access to trace elements present in deep-ocean seawater. In addition, hydrogen can be extracted from water through electrolysis for use in fuel cells.

OTEC is a relatively expensive technology, since the construction of costly OTEC plants and infrastructure is necessary before power can be generated. However, once facilities are made operational, it may be possible to generate relatively inexpensive electricity. Floating facilities may be more feasible than land-based ones, because the number of land-based sites with access to deep water in the tropics is limited.

Land, Shelf and Floating Sites

OTEC has the potential to produce gig watts of electrical power, and in conjunction with electrolysis, could produce enough hydrogen to completely replace all projected global fossil fuel consumption. Reducing costs remains an unsolved challenge, however. OTEC plants require a long, large diameter intake pipe, which is submerged a kilometer or more into the ocean's depths, to bring cold water to the surface.

Land Based

Land based and near-shore facilities offer three main advantages over those located in deep water. Plants constructed on or near land do not require sophisticated mooring, lengthy power cables, or the more extensive maintenance associated with open-ocean environments. They can be installed in sheltered areas so that they are relatively safe from storms and heavy seas. Electricity, desalinated water, and cold, nutrient-rich seawater could be transmitted from near-shore facilities via trestle bridges or causeways. In addition, land-based or near-shore sites allow plants to operate with related industries such as mariculture or those that require desalinated water.

Favored locations include those with narrow shelves (volcanic islands), steep (15-20 degrees) offshore slopes, and relatively smooth sea floors. These sites minimize the length of the intake pipe. A land-based plant could be built well inland from the shore, offering more protection from storms, or on the beach, where the pipes would be shorter. In either case, easy access for construction and operation helps lower costs.

Land-based or near-shore sites can also support mariculture or chilled water agriculture. Tanks or lagoons built on shore allow workers to monitor and control miniature marine environments. Mariculture products can be delivered to market via standard transport.

One disadvantage of land-based facilities arises from the turbulent wave action in the surf zone. OTEC discharge pipes should be placed in protective trenches to prevent subjecting them to extreme stress during storms and prolonged periods of heavy seas. Also, the mixed discharge of cold and warm seawater may need to be carried several hundred meters offshore to reach the proper depth before it is released, requiring additional expense in construction and maintenance.

One way that OTEC systems can avoid some of the problems and expenses of operating in a surf zone is by building them just offshore in waters ranging from 10 to 30 meters deep. This type of plant would use shorter (and therefore less costly) intake and discharge pipes, which would avoid the dangers of turbulent surf. The plant itself, however, would require protection from the marine environment, such as breakwaters and erosion-resistant foundations, and the plant output would need to be transmitted to shore.

Shelf Based

To avoid the turbulent surf zone as well as to move closer to the cold-water resource, OTEC plants can be mounted to the continental shelf at depths up to 100 meters (330 ft). A shelf-mounted plant could be towed to the site and affixed to the sea bottom. This type of construction is already used for offshore oil rigs. The complexities of operating an OTEC plant in deeper water may make them more expensive than land-based approaches. Problems include the stress of open-ocean conditions and more difficult product delivery. Addressing strong ocean currents and large waves adds engineering and construction expense. Platforms require extensive pilings to maintain a stable base. Power delivery can require long underwater cables to reach land. For these reasons, shelf-mounted plants are less attractive.

Floating

Floating OTEC facilities operate off-shore. Although potentially optimal for large systems, floating facilities present several difficulties. The difficulty of mooring plants in very deep water complicates power delivery. Cables attached to floating platforms are more susceptible to damage, especially during storms. Cables at depths greater than 1000 meters are difficult to maintain and repair. Riser cables, which connect the sea bed and the plant, need to be constructed to resist entanglement.

As with shelf-mounted plants, floating plants need a stable base for continuous operation. Major storms and heavy seas can break the vertically suspended cold-water pipe and interrupt warm water intake as well. To help prevent these problems, pipes can be made of flexible polyethylene attached to the bottom of the platform and gimballed with joints or collars. Pipes may need to be uncoupled from the plant to prevent storm damage. As an alternative to a warm-water pipe, surface water can be drawn directly into the platform; however, it is necessary to prevent the intake flow from being damaged or interrupted during violent motions caused by heavy seas.

Connecting a floating plant to power delivery cables requires the plant to remain relatively stationary. Mooring is an acceptable method, but current mooring technology is limited to depths of about 2,000 meters (6,600 ft). Even at shallower depths, the cost of mooring may be prohibitive.

Related Activities

OTEC has uses other than power production.

Desalination

Desalinated water can be produced in open- or hybrid-cycle plants using surface condensers to turn evaporated seawater into potable water. System analysis indicates that a 2-megawatt plant could produce about 4,300 cubic metres (150,000 cu ft) of desalinated water each day. Another system patented by Richard Bailey creates condensate water by regulating deep ocean water flow through surface condensers correlating with fluctuating dew-point temperatures. This condensation system uses no incremental energy and has no moving parts.

On March 22, 2015, Saga University opened a Flash-type desalination demonstration facility on Kumejima. This satellite of their Institute of Ocean Energy uses post-OTEC deep seawater from the Okinawa OTEC Demonstration Facility and raw surface seawater to produce desalinated water. Air is extracted from the closed system with a vacuum pump. When raw sea water is pumped into the flash chamber it boils, allowing pure steam to rise and the salt and remaining seawater to be removed. The steam is returned to liquid in a heat exchanger with cold post-OTEC deep seawater. The desalinated water can be used in hydrogen production or drinking water (if minerals are added).

Air Conditioning

The 41° F (5° C) cold seawater made available by an OTEC system creates an opportunity to provide large amounts of cooling to industries and homes near the plant. The water can be used in chilled-water coils to provide air-conditioning for buildings. It is estimated that a pipe 1 foot (0.30 m) in diameter can deliver 4,700 gallons of water per minute. Water at 43° F (6° C) could provide more than enough air-conditioning for a large building. Operating 8,000 hours per year in lieu of electrical conditioning selling for 5-10¢ per kilowatt-hour, it would save $200,000-$400,000 in energy bills annually.

The InterContinental Resort and Thalasso-Spa on the island of Bora Bora uses an SWAC system to air-condition its buildings. The system passes seawater through a heat exchanger where it cools freshwater in a closed loop system. This freshwater is then pumped to buildings and directly cools the air.

In 2010, Copenhagen Energy opened a district cooling plant in Copenhagen, Denmark. The plant delivers cold seawater to commercial and industrial buildings, and has reduced electricity consumption by 80 percent. Ocean Thermal Energy Corporation (OTE) has designed a 9800-ton SDC system for a vacation resort in The Bahamas.

Chilled Soil Agriculture

OTEC technology supports chilled-soil agriculture. When cold seawater flows through underground pipes, it chills the surrounding soil. The temperature difference between roots in the cool soil and leaves in the warm air allows plants that evolved in temperate climates to be grown in the subtropics. Dr. John P. Craven, Dr. Jack Davidson and Richard Bailey patented this process and demonstrated it at a research facility at the Natural Energy Laboratory of Hawaii Authority (NELHA). The research facility demonstrated that more than 100 different crops can be grown using this system. Many normally could not survive in Hawaii or at Keahole Point.

Japan has also been researching agricultural uses of Deep Sea Water since 2000 at the Okinawa Deep Sea Water Research Institute on Kume Island. The Kume Island facilities use regular water cooled by Deep Sea Water in a heat exchanger run through pipes in the ground to cool soil. Their techniques have developed an important resource for the island community as they now produce spinach, a winter vegetable, commercially year round. An expansion of the deep seawater agriculture facility was completed by Kumejima Town next to the OTEC Demonstration Facility in 2014. The new facility is for researching the economic practicality of chilled-soil agriculture on a larger scale.

Aquaculture

Aquaculture is the best-known byproduct, because it reduces the financial and energy costs of pumping large volumes of water from the deep ocean. Deep ocean water contains high concentrations of essential nutrients that are depleted in surface waters due to biological consumption. This "artificial upwelling" mimics the natural upwellings that are responsible for fertilizing and supporting the world's largest marine ecosystems, and the largest densities of life on the planet.

Cold-water delicacies, such as salmon and lobster, thrive in this nutrient-rich, deep, seawater. Microalgae such as *Spirulina*, a health food supplement, also can be cultivated. Deep-ocean water can be combined with surface water to deliver water at an optimal temperature.

Non-native species such as salmon, lobster, abalone, trout, oysters, and clams can be raised in pools supplied by OTEC-pumped water. This extends the variety of fresh seafood products available for nearby markets. Such low-cost refrigeration can be used to maintain the quality of harvested fish, which deteriorate quickly in warm tropical regions. In Kona, Hawaii, aquaculture companies working with NELHA generate about $40 million annually, a significant portion of Hawaii's GDP.

The NELHA plant established in 1993 produced an average of 7,000 gallons of freshwater per day. KOYO USA was established in 2002 to capitalize on this new economic opportunity. KOYO bottles the water produced by the NELHA plant in Hawaii. With the capacity to produce one million bottles of water every day, KOYO is now Hawaii's biggest exporter with $140 million in sales.

Hydrogen Production

Hydrogen can be produced via electrolysis using OTEC electricity. Generated steam with electrolyte compounds added to improve efficiency is a relatively pure medium for hydrogen production. OTEC can be scaled to generate large quantities of hydrogen. The main challenge is cost relative to other energy sources and fuels.

Mineral Extraction

The ocean contains 57 trace elements in salts and other forms and dissolved in solution. In the past, most economic analyses concluded that mining the ocean for trace elements would be unprofitable, in part because of the energy required to pump the water. Mining generally targets minerals that occur in high concentrations, and can be extracted easily, such as magnesium. With OTEC plants supplying water, the only cost is for extraction. The Japanese investigated the possibility of extracting uranium and found developments in other technologies (especially materials sciences) were improving the prospects.

Thermodynamics

A rigorous treatment of OTEC reveals that a 20 °C temperature difference will provide as much energy as a hydroelectric plant with 34 m head for the same volume of water flow. The low temperature difference means that water volumes must be very large to extract useful amounts of heat. A 100MW power plant would be expected to pump on the order of 12 million gallons (44,400 tonnes) per minute. For comparison, pumps must move a mass of water greater than the weight of the *battleship Bismarck*, which weighed 41,700 tonnes, every minute. This makes pumping a substantial parasitic drain on energy production in OTEC systems, with one Lockheed design consuming 19.55 MW in pumping costs for every 49.8 MW net electricity generated. For OTEC schemes using heat exchangers, to handle this volume of water the exchangers need to be enormous compared to those used in conventional thermal power generation plants, making them one of the most critical components due to their impact on overall efficiency. A 100 MW OTEC power plant would require 200 exchangers each larger than a 20-foot shipping container making them the single most expensive component.

Variation of ocean temperature with depth

The total insolation received by the oceans (covering 70% of the earth's surface, with clearness index of 0.5 and average energy retention of 15%) is:

5.45×10^{18} MJ/yr $\times 0.7 \times 0.5 \times 0.15 = 2.87 \times 10^{17}$ MJ/yr

We can use Beer–Lambert–Bouguer's law to quantify the solar energy absorption by water,

$$-\frac{dI(y)}{dy} = \mu I$$

Where, y is the depth of water, I is intensity and μ is the absorption coefficient. Solving the above differential equation,

$$I(y) = I_0 \exp(-\mu y)$$

The absorption coefficient μ may range from 0.05 m^{-1} for very clear fresh water to 0.5 m^{-1} for very salty water.

Since the intensity falls exponentially with depth y, heat absorption is concentrated at the top layers. Typically in the tropics, surface temperature values are in excess of 25° C (77° F), while at 1 kilometer (0.62 mi), the temperature is about 5–10 °C (41–50 °F).

The warmer (and hence lighter) waters at the surface means there are no thermal convection currents. Due to the small temperature gradients, heat transfer by conduction is too low to equalize the temperatures. The ocean is thus both a practically infinite heat source and a practically infinite heat sink.

This temperature difference varies with latitude and season, with the maximum in tropical, subtropical and equatorial waters. Hence the tropics are generally the best OTEC locations.

Open/Claude Cycle

In this scheme, warm surface water at around 27° C (81° F) enters an evaporator at pressure slightly below the saturation pressures causing it to vaporize.

$$H_1 = H_f$$

Where, H_f is enthalpy of liquid water at the inlet temperature, T_1.

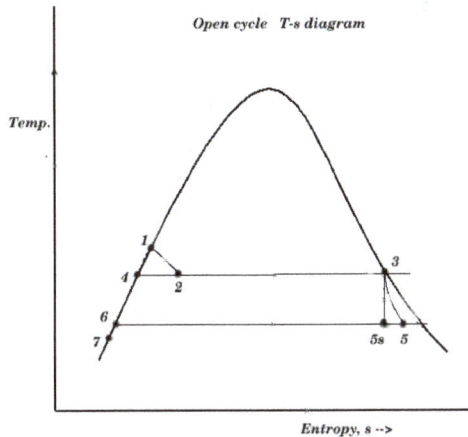

Open cycle T-s diagram

This temporarily superheated water undergoes volume boiling as opposed to pool boiling in conventional boilers where the heating surface is in contact. Thus the water partially flashes to steam with two-phase equilibrium prevailing. Suppose that the pressure inside the evaporator is maintained at the saturation pressure, T_2.

$$H_2 = H_1 = H_f + x_2 H_{fg}$$

Here, x_2 is the fraction of water by mass that vaporizes. The warm water mass flow rate per unit turbine mass flow rate is $1/x_2$.

The low pressure in the evaporator is maintained by a vacuum pump that also removes the dissolved non-condensable gases from the evaporator. The evaporator now contains a mixture of water and steam of very low vapor quality (steam content). The steam is separated from the water as saturated vapor. The remaining water is saturated and is discharged to the ocean in the open cycle. The steam is a low pressure/high specific volume working fluid. It expands in a special low pressure turbine.

$$H_3 = H_g$$

Here, H_g corresponds to T_2. For an ideal isentropic (reversible adiabatic) turbine,

$$s_{5,s} = s_3 = s_f + x_{5,s} s_{fg}$$

The above equation corresponds to the temperature at the exhaust of the turbine, T_5. $x_{5,s}$ is the mass fraction of vapor at state 5.

The enthalpy at T_5 is,

$$H_{5,s} = H_f + x_{5,s} H_{fg}$$

This enthalpy is lower. The adiabatic reversible turbine work = $H_3 - H_{5,s}$.

Actual turbine work $W_T = (H_3 - H_{5,s})$ x *polytrophic efficiency*

$$H_5 = H_3 - \text{actual work}$$

The condenser temperature and pressure are lower. Since the turbine exhaust is to be discharged back into the ocean, a direct contact condenser is used to mix the exhaust with cold water, which results in a near-saturated water. That water is now discharged back to the ocean.

$H_6 = H_p$ at T_5. T_7 is the temperature of the exhaust mixed with cold sea water, as the vapour content now is negligible,

$$H_7 \approx H_f \ at \ T_7$$

The temperature differences between stages include that between warm surface water and working steam, that between exhaust steam and cooling water, and that between cooling water reaching the condenser and deep water. These represent external irreversibilities that reduce the overall temperature difference.

The cold water flow rate per unit turbine mass flow rate,

$$\dot{m}_c = \frac{H_5 - H_6}{H_6 - H_7}$$

Turbine mass flow rate, $\dot{M}_T = \dfrac{\text{turbine work required}}{\dot{W}_T}$

Warm water mass flow rate, $\dot{M}_w = \dot{M}_T \, \dot{m}_w$

Cold water mass flow rate $\dot{M}_c = \dot{M}_T \dot{m}_c$

Closed Anderson Cycle

A Developed starting in the 1960s by J. Hilbert Anderson of Sea Solar Power, Inc. In this cycle, Q_H is the heat transferred in the evaporator from the warm sea water to the working fluid. The working fluid exits the evaporator as a gas near its dew point.

The high-pressure, high-temperature gas then is expanded in the turbine to yield turbine work, W_T. The working fluid is slightly superheated at the turbine exit and the turbine typically has an efficiency of 90% based on reversible, adiabatic expansion.

From the turbine exit, the working fluid enters the condenser where it rejects heat, $-Q_C$, to the cold sea water. The condensate is then compressed to the highest pressure in the cycle, requiring condensate pump work, W_C. Thus, the Anderson closed cycle is a Rankine-type cycle similar to the conventional power plant steam cycle except that in the Anderson cycle the working fluid is never superheated more than a few degrees Fahrenheit. Owing to viscous effects, working fluid pressure drops in both the evaporator and the condenser. This pressure drop, which depends on the types of heat exchangers used, must be considered in final design calculations but is ignored here to simplify the analysis. Thus, the parasitic condensate pump work, W_C, computed here will be lower than if the heat exchanger pressure drop was included. The major additional parasitic energy requirements in the OTEC plant are the cold water pump work, W_{CT}, and the warm water pump work, W_{HT}. Denoting all other parasitic energy requirements by W_A, the net work from the OTEC plant, W_{NP} is:

$$W_{NP} = W_T - W_C - W_{CT} - W_{HT} - W_A$$

The thermodynamic cycle undergone by the working fluid can be analyzed without detailed consideration of the parasitic energy requirements. From the first law of thermodynamics, the energy balance for the working fluid as the system is:

$$W_N = Q_H - Q_C$$

Where $W_N = W_T + W_C$ is the network for the thermodynamic cycle. For the idealized case in which there is no working fluid pressure drop in the heat exchangers,

$$Q_H = \int_H T_H ds$$

And

$$Q_C = \int_C T_C ds$$

So that the net thermodynamic cycle work becomes,

$$W_N = \int_H T_H ds - \int_C T_C ds$$

Subcooled liquid enters the evaporator. Due to the heat exchange with warm sea water, evaporation takes place and usually superheated vapor leaves the evaporator. This vapor drives the turbine and the 2-phase mixture enters the condenser. Usually, the subcooled liquid leaves the condenser and finally, this liquid is pumped to the evaporator completing a cycle.

Environmental Impact

Carbon dioxide dissolved in deep cold and high pressure layers is brought up to the surface and released as the water warms.

Mixing of deep ocean water with shallower water brings up nutrients and makes them available to shallow water life. This may be an advantage for aquaculture of commercially important species, but may also unbalance the ecological system around the power plant.

OTEC plants use very large flows of warm surface seawater and cold deep seawater to generate constant renewable power. The deep seawater is oxygen deficient and generally 20-40 times more nutrient rich (in nitrate and nitrite) than shallow seawater. When these plumes are mixed, they are slightly denser than the ambient seawater. Though no large scale physical environmental testing of OTEC has been done, computer models have been developed to simulate the effect of OTEC plants.

Hydrodynamic Modeling

In 2010, a computer model was developed to simulate the physical oceanographic effects of one or several 100 megawatt OTEC plant(s). The model suggests that OTEC plants can be configured such that the plant can conduct continuous operations, with resulting temperature and nutrient variations that are within naturally occurring levels. Studies to date suggest that by discharging the OTEC flows downwards at a depth below 70 meters, the dilution is adequate and nutrient enrichment is small enough so that 100-megawatt OTEC plants could be operated in a sustainable manner on a continuous basis.

Biological Modeling

The nutrients from an OTEC discharge could potentially cause increased biological activity if they accumulate in large quantities in the photic zone. In 2011 a biological component was added to the hydrodynamic computer model to simulate the biological response to plumes from 100 megawatt OTEC plants. In all cases modeled (discharge at 70 meters depth or more), no unnatural variations occurs in the upper 40 meters of the ocean's surface. The picoplankton response in the 110 - 70 meter depth layer is approximately a 10-25% increase, which is well within naturally occurring variability. The nanoplankton response is negligible. The enhanced productivity of diatoms (microplankton) is small. The subtle phytoplankton increase of the baseline OTEC plant suggests that higher-order biochemical effects will be very small.

Studies

A previous Final Environmental Impact Statement (EIS) for the United States' NOAA from 1981 is available, but needs to be brought up to current oceanographic and engineering standards. Studies have been done to propose the best environmental baseline monitoring practices, focusing on a set of ten chemical oceanographic parameters relevant to OTEC. Most recently, NOAA held an OTEC Workshop in 2010 and 2012 seeking to assess the physical, chemical, and biological impacts and risks, and identify information gaps or needs.

The Tethys database provides access to scientific literature and general information on the potential environmental effects of OTEC.

Technical Difficulties

Dissolved Gases

The performance of direct contact heat exchangers operating at typical OTEC boundary conditions is important to the Claude cycle. Many early Claude cycle designs used a surface condenser since their performance was well understood. However, direct contact condensers offer significant disadvantages. As cold water rises in the intake pipe, the pressure decreases to the point where gas begins to evolve. If a significant amount of gas comes out of solution, placing a gas trap before the direct contact heat exchangers may be justified. Experiments simulating conditions in the warm water intake pipe indicated about 30% of the dissolved gas evolves in the top 8.5 meters (28 ft) of the tube. The trade-off between pre-dearation of the seawater and expulsion of non-condensable gases from the condenser is dependent on the gas evolution dynamics, deaerator efficiency, head loss, vent compressor efficiency and parasitic power. Experimental results indicate vertical spout condensers perform some 30% better than falling jet types.

Microbial Fouling

Because raw seawater must pass through the heat exchanger, care must be taken to maintain good thermal conductivity. Biofouling layers as thin as 25 to 50 micrometres (0.00098 to 0.00197 in) can degrade heat exchanger performance by as much as 50%. A 1977 study in which mock heat exchangers were exposed to seawater for ten weeks concluded that although the level of microbial fouling was low, the thermal conductivity of the system was significantly impaired. The apparent discrepancy between the level of fouling and the heat transfer impairment is the result of a thin layer of water trapped by the microbial growth on the surface of the heat exchanger.

Another study concluded that fouling degrades performance over time, and determined that although regular brushing was able to remove most of the microbial layer, over time a tougher layer formed that could not be removed through simple brushing. The study passed sponge rubber balls through the system. It concluded that although the ball treatment decreased the fouling rate it was not enough to completely halt growth and brushing was occasionally necessary to restore capacity. The microbes regrew more quickly later in the experiment (i.e. brushing became necessary more often) replicating the results of a previous study. The increased growth rate after subsequent cleanings appears to result from selection pressure on the microbial colony.

Continuous use of 1 hour per day and intermittent periods of free fouling and then chlorination periods (again 1 hour per day) were studied. Chlorination slowed but did not stop microbial growth; however chlorination levels of .1 mg per liter for 1 hour per day may prove effective for long term operation of a plant. The study concluded that although microbial fouling was an issue for the warm surface water heat exchanger, the cold water heat exchanger suffered little or no bio fouling and only minimal inorganic fouling.

Besides water temperature, microbial fouling also depends on nutrient levels, with growth occurring faster in nutrient rich water. The fouling rate also depends on the material used to construct the heat exchanger. Aluminium tubing slows the growth of microbial life, although the oxide layer which forms on the inside of the pipes complicates cleaning and leads to larger efficiency losses. In contrast, titanium tubing allows bio fouling to occur faster but cleaning is more effective than with aluminium.

Sealing

The evaporator, turbine, and condenser operate in partial vacuum ranging from 3% to 1% of atmospheric pressure. The system must be carefully sealed to prevent in-leakage of atmospheric air that can degrade or shut down operation. In closed-cycle OTEC, the specific volume of low-pressure steam is very large compared to that of the pressurized working fluid. Components must have large flow areas to ensure steam velocities do not attain excessively high values.

Parasitic Power Consumption by Exhaust Compressor

An approach for reducing the exhaust compressor parasitic power loss is as follows. After most of the steam has been condensed by spout condensers, the non-condensible gas steam mixture is passed through a counter current region which increases the gas-steam reaction by a factor of five. The result is an 80% reduction in the exhaust pumping power requirements.

Cold Air/Warm Water Conversion

In winter in coastal Arctic locations, the delta T between the seawater and ambient air can be as high as 40° C (72° F). Closed-cycle systems could exploit the air-water temperature difference. Eliminating seawater extraction pipes might make a system based on this concept less expensive than OTEC. This technology is due to H. Barjot, who suggested butane as cryogen, because of its boiling point of −0.5° C (31.1° F) and its non-solubility in water. Assuming a level of efficiency of realistic 4%, calculations show that the amount of energy generated with one cubic meter water at a temperature of 2° C (36° F) in a place with an air temperature of −22° C (−8° F) equals the amount of energy generated by letting this cubic meter water run through a hydroelectric plant of 4000 feet (1,200 m) height.

Barjot Polar Power Plants could be located on islands in the polar region or designed as swimming barges or platforms attached to the ice cap. The weather station Myggbuka at Greenlands east coast for example, which is only 2,100 km away from Glasgow, detects monthly mean temperatures below −15° C (5° F) during 6 winter months in the year.

Seawater Air-conditioning

Seawater Air Conditioning (SWAC) is an alternate-energy system that uses the cold water from the deep ocean (and in some cases a deep lake) to cool buildings.

Along many ocean coastlines and lake shorelines, there is reasonable access to naturally cold water that is as cold or colder than the water used in conventional air conditioning systems. If this water can be tapped, then the significant power for operating mechanical chillers to keep the chilled water cold can be eliminated.

The adjacent temperature profile illustrates the temperature vs depth that is typical for the world's tropical deep oceans in the summertime: 7° C or below can be reached at 700m depth, 5° C. or below at 1000m. The deep-water portion of this profile changes little seasonally and therefore cold water is available on a year round basis. In more northern climates, very cold water can be reached at shallower depths during the summer – in both oceans and in deep water lakes.

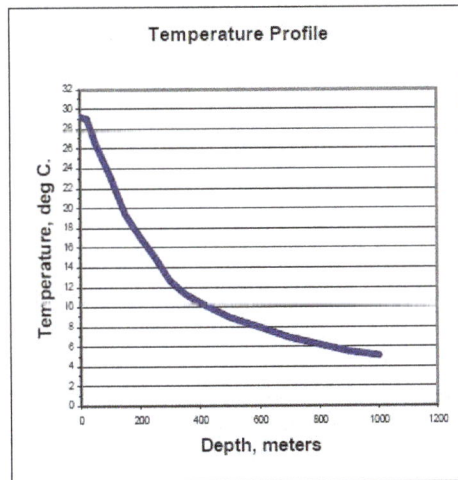

Temperature Profile

The basic concept of seawater air conditioning is to take advantage of available deep cold seawater to cool the chilled water in one or more buildings as opposed to using more energy intensive refrigeration systems.

A seawater air conditioning system is illustrated below. The buildings to the far right are identical internally to buildings cooled with conventional A/C. Chilled fresh water moves through these buildings with the same temperatures and flows of conventional systems. The seawater and chilled water pumps and heat exchangers would typically be located at the shoreline in a cooling station.

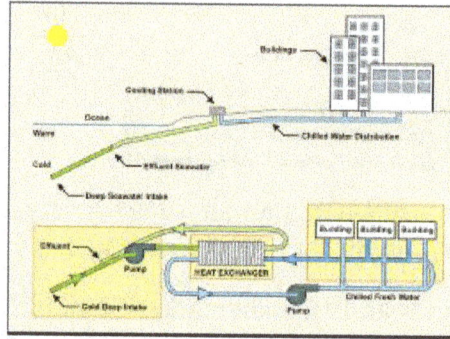

The main components of a seawater air conditioning system are the seawater supply system, the heat exchanger or cooling station and the fresh water distribution system. These basic components can be optimized for each specific location, climate and building.

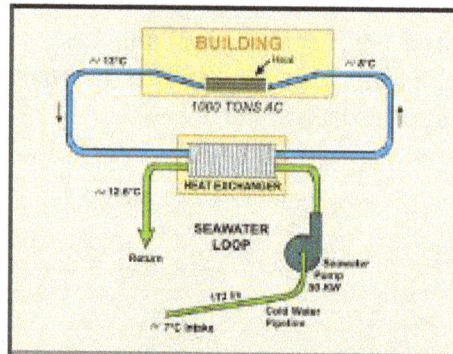

This schematic is an alternate view of a basic centralized seawater air conditioning system. The chilled water loop is fresh water and operating at the same temperatures as with conventional AC.

The interior of the building is unchanged with SWAC systems. The chilled water is kept cool through a heat exchanger with a counter flow of deep cold seawater. The heat exchanger is titanium to eliminate corrosion and fouling does not occur because of the purity of the deep seawater. Seawater is brought to the site through a deep-water polyethylene pipeline. These pipelines reach out several kilometers offshore and have a nominal intake depth of 700 m. The effluent seawater is discharged though a second pipe at a depth of approximately 40 m.

Seawater air conditioning is not technically complex nor is it a high technical risk. It is established technology being applied in an innovative way. All the components necessary exist and have been operated under the conditions required.

Use of an Auxiliary Chiller: In some cases, it is either too costly or impractical to Supply seawater at the necessary low temperatures to maintain minimum temperatures in the chilled water loop. The distance offshore to reach sufficiently cold water might be prohibitive or the ocean depth may simply not be available. It is sometimes economically possible to use auxiliary chillers to supplement the cooling provided by the seawater exposure. This is illustrated below. The fresh chill water is first cooled by seawater through a heat exchanger and then secondarily cooled with an auxiliary chiller. The auxiliary chiller is basically a refrigeration system with its condenser cooled by the returning flow of cool seawater. With the condenser kept cool, the auxiliary chiller can operate at an extremely high efficiency – as high as double that of a conventional chiller.

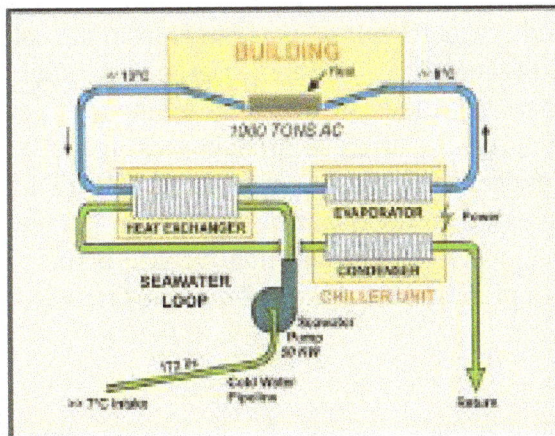

Cold Storage: A SWAC system has a high capital cost and a low operating cost. The peak capacity of the system must match the peak demand of the buildings that it serves. These demands are not constant throughout the day or throughout the year, and the total system is frequently not being used to its maximum capacity. Therefore, capital dollars are spent on a system that may not always be used to its maximum potential. A means of minimizing the capital cost is to use cold-water storage. The seawater air conditioning system would be operated 100 percent of the time and when the building demands are low, the excess capacity is directed into a storage system of cold fresh water. When A/C demand is at its peak, the cold water is drained from its storage to meet the demand.

Cold water storage tanks are commercially available that are constant volume; the warm water remains at the top and the coldest water remains at the bottom. These tanks are now used in conjunction with conventional A/C systems to take advantage of low, off-peak electrical rates.

Environmental Aspects

A SWAC system has significant environmental benefits: These include drastic reductions in electricity consumption which reduces air pollution and greenhouse gas production, and substitution of simple heat exchangers for chiller machinery which often use ozone-depleting chlorofluorocarbons (CFCs).

The existence of the deep ocean heat sink results from natural climatic processes where water is cooled at the poles, becomes dense and sinks to deeper water and slowly moves toward the equator. The cold ocean is therefore both immense and renewable.

Return water from a SWAC system can be handled in a number of ways. Typically it is returned to the ocean at a location where the return water temperature nearly matches the ambient water.

There are significant secondary applications for this seawater. Secondary cooling, aquaculture, desalination and even agriculture can benefit from the cold seawater. Aquaculturists value the water because it is clean and disease free. When used in conjunction with a warm source of water, they can have any temperature seawater their product needs. Secondary cooling can be used in greenhouses and other locations where humidity control is not a major factor. Finally, research in Hawaii has shown that even an arid land can be made highly productive with low fresh water consumption by cooling the soil and the roots of many tropical and non-tropical plants. Deep seawater is also desalinated and sold as a premium drinking water in the orient.

Economic Viability

The economic viability of a SWAC system is site specific. Each location has unique opportunities as well as problems. The main factors influencing the economic viability of a specific location include:

- *The distance offshore to cold water*: shorter pipelines are more economical than long pipelines.

- *The size of the air conditioning load:* there is an economy of scale associated with SWAC – systems less than 1000 tons are more difficult to justify economically.

- *The percent utilization of the air conditioning system:* The higher the utilization throughout the year, the higher the direct benefits.

- *The local cost of electricity*: A high cost of electricity makes conventional AC more costly and SWAC, in comparison, more attractive. Any cost analysis should include current and future costs of electricity.

- *The complexity of the distribution system on shore:* SWAC works best with a district cooling arrangement, where many buildings are cooled taking.

Advantage of The Economy of Scale

SWAC is even more economical if this distribution system is compact.

The adjacent figure illustrates the difference in lifetime costs for a conventional AC system and a typical SWAC system. The costs are broken down into capital, operating (energy) and maintenance. The primary cost of a SWAC system is in the initial capital cost. The operating and maintenance costs are small. For a conventional AC system, the primary cost is in the power consumed over its lifetime. Hence, SWAC systems are ideal for base load AC that has high utilization and conventional AC may be better for situations of infrequent use.

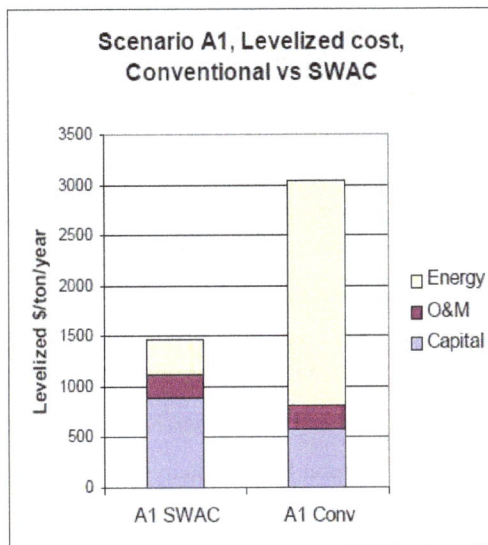

It's important to note that there is a dramatic economy of scale as the size of the pipeline increases. The reason is that the cold water pipe costs per liter of water delivered decreases as the pipeline size increases and temperature rise via large pipelines is practically negligible. The figure to the right illustrates five SWAC scenarios of varying overall size; the two bars compare the life time cost difference between conventional AC and SWAC.

Makai has performed SWAC feasibility studies of a variety of sites. Typical results are that electrical consumption is reduced by 80 to 90 percent. Simple payback can be from three to seven years, and long term costs can be half that of a conventional air conditioning system. Not all locations, however, are ideal. Some have poor access to deep cold-water sources or the overall size is too small to be economical. Each site is unique.

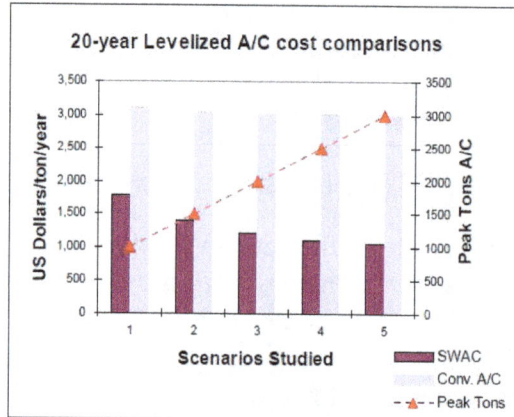

20-year Levelized A/C cost comparisons

References

- Takahashi, Masayuki Mac; Translated by: Kitazawa, Kazuhiro and Snowden, Paul (2000) [1991]. Deep Ocean Water as Our Next Natural Resource. Tokyo, Japan: Terra Scientific Publishing Company. ISBN 978-4-88704-125-7

- Dam: newworldencyclopedia.org, Retrieved 25 June 2018

- Mitsui, T.; Ito, F.; Seya, Y.; Nakamoto, Y. (September 1983). "Outline of the 100 kW OTEC Pilot Plant in the Republic of Nauru". IEEE Transactions on Power Apparatus and Systems. PAS-102 (9): 3167–3171. Bibcode:1983ITPAS.102.3167M. doi:10.1109/TPAS.1983.318124

- Challenges, Wave-energy-converters: theseusproject.eu, Retrieved 23 May 2018

- L. Meyer; D. Cooper; R. Varley. "Are We There Yet? A Developer's Roadmap to OTEC Commercialization" (PDF). Hawaii National Marine Renewable Energy Center. Retrieved 28 March 2013

- Ocean-thermal-energy-conversion: britannica.com, Retrieved 12 April 2018

- Shah, Yatish T. (2018-01-12). Thermal Energy: Sources, Recovery, and Applications. CRC Press. ISBN 9781315305936

- Liping Liu. "Feasibility of large-scale power plants based on thermoelectric effects." New Journal of Physics. doi:10.1088/1367-2630/16/12/123019

6

Applications of Hydropower

Hydropower is considered a means for economic development. Power generated using hydropower has a number of different applications. This chapter explores the use of hydroelectricity for flood risk management, in irrigation, business, etc.

Generating Electricity

Hydroelectricity is one of the most important sources of energy in the world. Hydroelectricity is one of the cheapest and non-polluting sources of power. Though it can cause ecological damage initially it has better climate compatibility than other major forms of energy like nuclear, coal, gas and others.

In order to generate electricity from the kinetic energy in moving water, the water has to move with sufficient speed and volume to spin a propeller-like device called a turbine, which in turn rotates a generator to generate electricity. Roughly speaking, one gallon of water per second falling one hundred feet can generate one kilowatt of electricity.

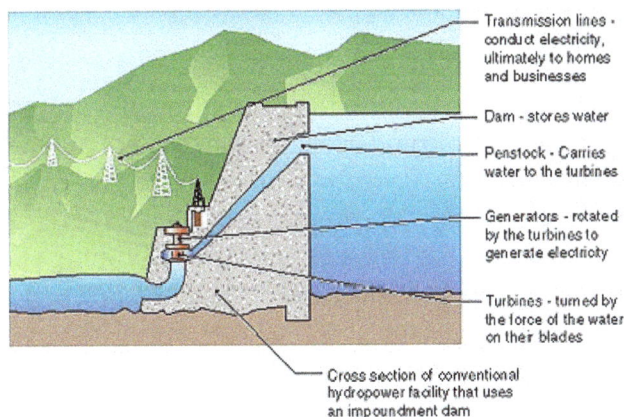

Transmission lines - conduct electricity, ultimately to homes and businesses

Dam - stores water

Penstock - Carries water to the turbines

Generators - rotated by the turbines to generate electricity

Turbines - turned by the force of the water on their blades

Cross section of conventional hydropower facility that uses an impoundment dam

To increase the volume of moving water, impoundments or dams are used to collect the water. An opening in the dam uses gravity to drop water down a pipe called a penstock. The moving water causes the turbine to spin, which causes magnets inside a generator to rotate and create electricity.

There are a variety of types of turbines used at hydropower facilities, and their use depends on the amount of hydraulic head (vertical distance between the dam and the turbine) at the plant. The most common are Kaplan, Francis, and Pelton wheel designs. Some of these designs, called reaction and impulse wheels, use not just the kinetic force of the moving water but also the water pressure.

The Kaplan turbine is similar to a boat propeller, with a runner (the turning part of a turbine) that has three to six blades, and can provide up to 400 MW of power. The Kaplan turbine is differentiated from other kinds of hydropower turbines because its performance can be improved by changing the pitch of the blades. The Francis turbine has a runner with nine or more fixed vanes. In this turbine design, which can be up to 800 MW in size, the runner blades direct the water so that it moves in an axial flow. The Pelton turbine consists of a set of specially shaped buckets that are mounted on the outside of a circular disc, making it look similar to a water wheel. Pelton turbines are typically used in high hydraulic head sites and can be as large as 200 MW.

Generators at Hoover Dam

Hydropower can also be generated without a dam, through a process known as run-of-the-river. In this case, the volume and speed of water is not augmented by a dam. Instead, a run-of-river project spins the turbine blades by capturing the kinetic energy of the moving water in the river. Hydropower projects that have dams can control when electricity is generated because the dams can control the timing and flow of the water reaching the turbines. Therefore these projects can choose to generate power when it is most needed and most valuable to the grid. Because run-of-river projects do not store water behind dams, they have much less ability to control the amount and timing of when electricity is generated.

Another type of hydropower technology is called pumped storage. In a pumped storage plant, water is pumped from a lower reservoir to a higher reservoir during off-peak times when electricity is relatively cheap, using electricity generated from other types of energy sources. Pumping the water uphill creates the potential to generate hydropower later on. When the hydropower power is needed, it is released back into the lower reservoir through turbines. Inevitably, some power is lost, but pumped storage systems can be up to 80 percent efficient.

Financial Benefits of Hydroelectricity

Alongside the obvious ecological benefits, there are various ways in which your hydro-electricity generation system can earn and save you money. It's one of the main reasons that a lot of commercial concerns are beginning to turn to the technology as a viable, and profitable alternative to simply taking power off the grid.

Hydroelectricity Investment Information

It is difficult to calculate the precise revenues that can be achieved given the large range of factors, such as the size and planning and development cost of the installation, and the fact that each project owner will use their energy in different ways.

The table below presents a rough estimate of how much revenue you can expect your hydro system to generate annually as well as when you can hope to break even:

System power	Initial system cost	Approx. annual income (less running costs)	Approx. breakeven point
5kW	£50,000	£5,000	10 years
25kW	£200,000	£25,000	8 years
50kW	£350,000	£45,000	7 years, 8 months
100kW	£500,000	£95,000	5 years, 3 months
250kW	£1,000,000	£160,000	6 years, 3 months*
1MW	£3,000,000	£450,000	6 years, 8 months

The apparent anomaly in breakeven points here can be attributed to the lower FiT rate available to larger systems.

The above table do not take into account potential savings from using the energy produced in your own home(s) or business. The total saving could therefore be potentially much higher than these estimates.

As noted already, the energy return from hydroelectricity systems is much more calculable than it is for other sources of renewable energy such as solar PV, as it is less dependent on variations in the local weather.

It is difficult to determine the potential revenue offered by tidal hydro systems given their scarcity.

Feed-in Tariffs for Hydroelectricity

Feed-in Tariffs (FiTs) are government subsidies which allow you to receive money for energy generated by your renewable energy installation that is fed to the National Grid. This is referred to as the generation tariff.

Along with lower energy bills, FiTs present the principal financial incentive for those who install green energy systems in homes, organisations, communities and businesses. FiTs allow to you receive money (tax free for non-business installations) from your energy supplier for every kilowatt hour (kWh) of electricity generated by your low-carbon energy installation.

FiTs also provide income for every unit of surplus energy that you do not use and export back to the grid. This is referred to as the export tariff. Off-grid systems are not eligible for the FiT export tariff, though they may still claim the generation tariff. For projects such as community installations, none of the electricity produced is fed directly into local homes but instead is exported to the grid and the resulting profit is then fed back into the community.

NB. For businesses, FiT payments count as income and will be taxed accordingly.

Despite their somewhat misleading name, Feed-in-Tariffs pay you for *all of the energy that the hydroelectric system produces*, whether you consume it yourself or sell it back onto the national grid.

The current FiT generation rates (valid until 31st March 2014) for hydroelectricity generation are as follows:

System output/capacity	Feed-in generation rate tariff
<15kW	21.65p/kWh
15kW-100kW	20.21p/kWh
100kW-500kW	15.98p/kWh
500kW-2MW	12.48p/kWh
2MW-5MW	3.23p/kWh

The export tariff for all system sizes (and indeed all eligible renewable energy generation installations) is 4.64p/kWh.

The rate at which energy is exported back onto the grid is dependent upon your installation's output and your energy consumption.

NB - Feed-in Tariffs are not currently available in Northern Ireland.

FiT rates are subject to change and only apply to systems installed before 31st March 2014. Tariffs are reviewed periodically by the Department for Energy and Climate change (DECC), but are locked in for a period of 20 years (the tariff 'lifetime') from the installation of the hydro system. FiT rates are also linked to the Retail Price Index (RPI), so they will rise with inflation.

Renewables Levy Exemption Certificates

Another way of saving money with renewable technology for businesses is to obtain a Renewables Levy Exemption Certificate (Renewables LEC).

Renewables LECs bypass the need to pay the Climate Change Levy (CCL), which is a tax on energy products (electricity, natural gas, LPG, coal) used in businesses. Businesses which use these energy sources currently pay the CCL at a rate of £5.24/MWh used.

If you use the energy produced by a renewable system (provided that the system capacity is between 50kW and 5MW) in your business, you may be eligible for an LEC, saving you the £5.24/MWh CCL. It is also possible to sell the certificates back to energy suppliers.

Charities and non-business organisations do not need to pay the Climate Change Levy.

Selling Shares for Community Hydroelectricity

Where community schemes are starting to benefit by hydro projects is when local people can buy shares which then yield an individual profit in addition to the profits made from the FiT that is fed back into the community.

Flood Risk Management

Hydro power plants can be used to regulate and reduce the risk of flooding. However, due to the large quantities of water stored in the dam of the Hydro power plants, they also bring with them a new risk of flooding.

The Impacts of Hydro Power Plants on Flooding

Hydro Power Plants Causing Flooding

The United Nations (UN) estimates that over 60% of the world's largest river systems have been disturbed by dams and other man-made diversions, signifying a huge impact on local ecology and the water cycle within the area. Some of these effects may include loss of habitat in the catchment area (both for communities and animals), plain flooding and subsequent loss of endemic vegetation, and deviation in fish population.

Hydro power plants can cause an increased risk in flooding. The plain flooding caused by hydro power plants are controlled to form the power plant's reservoir dam. The issue of uncontrolled flooding in relation to hydro power plants however are often due to a build-up of sediment, Further studies confirm this, stating that sedimentation in the reservoir leads to a reduction in the available reservoir volume. Thus the reservoir is no longer able to control large inflows—such as those brought on by unusually heavy rainfall. This can then lead to unforeseen discharge/overflowing of the dam.

Events like these demand different methods of mitigation. An example of non-structural amendments includes more conservative level control by gradual discharge through spillways, implemented in parallel with sedimentation monitoring. Additionally, structural methods such as reservoir deepening or dredging to increase retention volume, and longer operation times can prevent flood risk.Alternative methods include altering the discharge flow to be dissimilar to rainfall or inflow times and mobilising the sediment build-up.

Hydro Power Plants Preventing Flooding

However, often hydro power plants are actually used to control and prevent floods. The accumulation of such large volumes of water in the reservoir make flood management a crucial part of hydro power plants. Whereas undammed water systems would require observation of several waterways, damming allows for the maintenance and monitoring of a single body of water. Even the smallest of retention volumes reduce the likeliness of flooding. For example, the Kaprun Ache, in Austria, had a natural probity of experiencing a flood every 10 years, however now, after the construction of a hydro power plants with a reservoir, the probability of a major flood occurring has dropped to once every 75 years.

In Brazil hydro power plants, in conjunction with a matrix based monitoring system, are also used to control flooding. The system is based on a waiting volume in the reservoir, a volume of water retained to allow optimal energy production, but also to safeguard from potential flood-inflow. This statistic is calculated using mathematical models of recurrence time, flood probability, inflow period characteristics, operational constraints and other factors. These are all considered in the planning stage of the hydro power plants to ensure that any hydro power plants built can control potential flood events.

Micro Hydro Power Plants

Micro and mini hydro power schemes are often run of-the river, which do not directly interfere with the natural flow of the river, or they have small dams/wires/ catchment area to store some water above the turbine. Due to the small scale of the power plants they are general considered less environmentally intrusive than large scale hydro power plants. However, some studies have found that, some studies have found the micro hydro power sites increase the risk of flooding in an area. This is because the power site introduces a barrier into the river system, which can lead to a build-up of sediment which in turn can result in an increased risk of localised flooding. This is especially the case for micro hydro schemes which include a small catchment area.

However, micro and mini hydro power plant can also be designed so that they contribute to flood prevention, for example by providing additional flood plain storage.

Enabling Irrigation

Reservoirs are generally built with multiple functions in mind, and irrigation and hydroelectricity generation are often the main functions. The construction and operation of reservoirs has led to tradeoffs (negative relationships) and complementarities (or positive synergies) among these functions within the food-energy-water (FEW) nexus. Reservoirs or dams with both hydropower and irrigation functions buffer the fluctuations of natural stream flow and can provide reliable water supply for irrigation during dry periods. At the same time, reservoirs are operated to store water to build up a hydraulic head and then release water to generate hydroelectricity. When water stored in reservoirs is reserved for a future irrigation season, the elevated hydraulic head would increase hydroelectricity generation; similarly, water released for irrigation may reduce reservoir storage, thereby reducing hydroelectricity generation, especially during dry and hot periods, when demand for irrigation and energy might be largest.

Response of Hydroelectricity Generation to Irrigation Water use

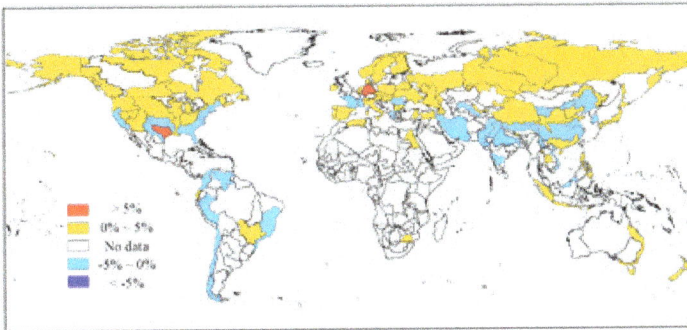

Figure: The sensitivity of hydropower generation to irrigation

The response of hydroelectricity generation to climatic inputs and irrigation water use is obtained through sensitivity analysis for the trained SVMs. Each independent variable (i.e. precipitation, potential evaporation and irrigation water use) is perturbed by one standard deviation around its mean value, and the corresponding change of the hydroelectricity generation is compared to that calculated from the mean value of the perturbing variable. Figure presents the sensitivity of hydroelectricity generation to irrigation. The FPUs with positive sensitivity values (i.e. in red and yellow color depicting complementary hydroelectricity- irrigation relationships) are mostly located in Southeast Asia, the Yellow and Yangtze rivers of China, the East and West Coast of the U.S., Russia, Canada and FPUs along the Andes. That is, storage for hydroelectricity generation in these FPUs enhances water supply for irrigation. Reservoirs constructed for hydroelectricity regulate stream flow, which benefits irrigation. Moreover, water stored during the wet season for irrigation in the dry season elevates the water head in reservoirs, leading to larger hydroelectricity generation, and the timing of the two is complementary, i.e. demands for hydroelectricity and for irrigation are complementing each other. Thus the constructed infrastructure and its operation contribute to the complementarity between hydroelectricity and irrigation.

FPUs with negative sensitivity values (i.e. in blue color showing competing hydroelectricity-irrigation relationships) are mostly located in regions such as the Central U.S., Northern Europe, India, Central Asia and Oceania. In these FPUs, increased hydropower coincides with reduced irrigation. Some of these regions have limited stream flow, especially during the irrigation season. In other regions, timing between demands for irrigation and hydropower releases differs. For example, an upstream region or country holds water in the summer irrigation season for hydroelectricity generation during the following winter.

Given the limited water resources in these regions, the tradeoff relationship between irrigation and hydropower may be mitigated through changes in food and energy trade with other regions.

In some regions, profits from food production can also be used to compensate for losses in hydroelectricity generation. Although hydroelectricity generation does not result in large consumptive water use (except for sreservoir evaporative losses and seepage in some places), irrigation water access of downstream regions may suffer from stream flow regulation through reservoir operations. In other regions, reservoir storage capacity may not be large enough to properly regulate stream flow such as to meet both hydroelectricity and irrigation demand. Reservoir operations in those regions have to partition water stored in the reservoir between hydropower and irrigation, and deal with the tradeoffs between the two functions. In these places, reservoir construction might help to alleviate these tradeoffs. In yet other places, storage augmentation might not address tradeoffs, at least not during climate extremes.

Permissions

Index

www.ingramcontent.com/pod-product-compliance
Lightning Source LLC
Chambersburg PA
CBHW061936190326
41458CB00009B/2751